T0122331

Studies in Systems, Decision and Control

Volume 114

Series editor

Janusz Kacprzyk, Polish Academy of Sciences, Warsaw, Poland
e-mail: kacprzyk@ibspan.waw.pl

About this Series

The series "Studies in Systems, Decision and Control" (SSDC) covers both new developments and advances, as well as the state of the art, in the various areas of broadly perceived systems, decision making and control- quickly, up to date and with a high quality. The intent is to cover the theory, applications, and perspectives on the state of the art and future developments relevant to systems, decision making, control, complex processes and related areas, as embedded in the fields of engineering, computer science, physics, economics, social and life sciences, as well as the paradigms and methodologies behind them. The series contains monographs, textbooks, lecture notes and edited volumes in systems, decision making and control spanning the areas of Cyber-Physical Systems, Autonomous Systems, Sensor Networks, Control Systems, Energy Systems, Automotive Systems, Biological Systems, Vehicular Networking and Connected Vehicles, Aerospace Systems, Automation, Manufacturing, Smart Grids, Nonlinear Systems, Power Systems, Robotics, Social Systems, Economic Systems and other. Of particular value to both the contributors and the readership are the short publication timeframe and the world-wide distribution and exposure which enable both a wide and rapid dissemination of research output.

More information about this series at http://www.springer.com/series/13304

Kofi Kissi Dompere

The Theory of Info-Dynamics: Rational Foundations of Information-Knowledge Dynamics

Kofi Kissi Dompere
Department of Economics
Howard University
Washington, DC
USA

ISSN 2198-4182 ISSN 2198-4190 (electronic)
Studies in Systems, Decision and Control
ISBN 978-3-319-87654-2 ISBN 978-3-319-63853-9 (eBook)
https://doi.org/10.1007/978-3-319-63853-9

Printed on acid-free paper

This Springer imprint is published by Springer Nature
The registered company is Springer International Publishing AG
The registered company address is: Gewerbestrasse 11, 6330 Cham, Switzerland

To my parents and the members of my extended family.

To those working to expand the common heritage of the stock of human knowledge without prejudice.

To all the authors in the reference list and those who are not referenced but whose ideas and efforts have influenced my cognitive development.

As well as expanded my epistemic model of the universal existence in one way or the other through information.

To all my teachers of my primary and secondary education in Ghana and my tertiary education at Temple University, Philadelphia in the United States of America.

Finally, to all those whose believe that there is a general theory of unified sciences on the basis of information and its applications.

Through a unified science, technology, engineering, and the families of ordinary and abstract languages on the basis of knowledge from the general theory.

Preamble

"At the beginning of his studies in natural sciences a student sincerely believes that such fundamental concepts as time, entropy, temperature and irreversibility are well defined and well understood by the experts, or at the lest, by the well-known authorities in thermodynamics, statistical mechanics and related subjects. It is, however, highly perplexing and embarrassing for a thought student to slowly discover later, that one after, the experts and the well-known authorities have never been really certain about the origin and meaning of the very fundamental concepts embedded in their own theories. Furthermore, often the authorities are in serious disagreement with each other (and with other views expressed by their colleagues), as to the proper definitions, origin, validity, and meaning of the aforementioned concepts. Sooner or later a person seeking a more consistent and universal answer to these questions finds that any respective answer requires endless (but seamless) links between philosophy of and the more earthly problems of classical thermodynamics, statistical mechanics, electromagnetism, relativity, and even cosmology!"

"Such a searching course into the open and unknown problems of today, requires the questioning, re-examination, screening, and, at times, radical rejection of "accepted" or "established" theories of past or present authorities. It criticizes and reprobates any traditional teaching, which eventually leads to inconsistencies or paradoxes, no matter how sacrosanct the "theorem" may be. It defends and preserves no argument which does not, by its own logic, consistency, and universality, defend and preserve itself. Such was always the course situated under the crisis of every science including thermodynamics" [Gal-Or, R8.29, p.211].

"The modern concept of information has considerably influenced the philosophical problems and has some important consequences for epistemology and for ontology, too. Information theory and cybernetics threw fresh light on such concepts as knowledge, time, evolution, etc.

If we want to illuminate the essence of the information principle in human cognition we must investigate the epistemological characteristics in general, we must determine the so-called epistemological situation, the situation of the human observer in the universe and their parameters. An analysis of this situation provides the basic conditions of cognition which are actually the limiting conditions.

The epistemological situation cannot be described in a wholly exhaustive manner—it is necessary to reduce it to its basic parameter, to construct its model. Each formulation is made in some language which uses certain concepts, principles, and relations; some of them are basic, starting. A certain description differs according to the initial presumptions; it even differs in the individual sciences. The task of philosophy is to try to give a general universal description of the epistemological situation, to determine its parameters in general and thereby to establish actually the relative limitation of cognition, the relative degree of our lack of knowledge—for purpose of overcoming it in then further development. Information theory brings new stimuli to this investigation. The human observes lives in the world which has certain space-time and other characteristics. He finds himself in a certain area of epistemological accessibility from whose parameters he must begin his investigation. He lives in the macro-world—between the megaword and microworld, he lives in a sphere where certain laws are valid, he is relatively limited by his senses, intellectual abilities, experience—he is limited by the properties of his environment and by his internal properties too. His empirical observation and logical thinking do not have unlimited precision. The totality of human knowledge is based on the finite number of acts of observation and thinking, the amount of information contained in them is finite. Human knowledge has a microscopic and statistical character since it proceeds from the biological and psychical attributes of man and has a particular degree of uncertainty (imprecision). Knowledge is dependent on the human information capacity: human information can never exceed the capacity of the epistemological channel although it can increase with the development of cognition" [Jiři Zeman, R8.29, pp. 245–246].

"Now this is the peculiarity of scientific method, that when once it has become a habit, that mind coverts all facts whatever into science. The field of science is unlimited; its material is endless, every group of natural phenomena, every phase of social life, every stage of past or present development is material for science. "The unity of all science consists alone in its method, not in its material". The man who classifies facts of any kind whatever, who sees their mutual relation, and describes their sequence, is applying the scientific method and is a man of science. The facts may belong to the past history of mankind, to the social statistics of our great cities, to the atmosphere of the most distant stars, to the digestive organs of a worm, or to the life of scarce bacillus. It is not the facts themselves which form science, but the methods by which they are dealt with". [K. Pearson, The Grammar of Science, London, Everyman Edition, 1938, p. 16]

"The need for understanding and managing natural and social phenomena resulted throughout history in man's constant preoccupation for representing, initially in a simplified form, then in an increasingly complex one, the surrounding realities. By synthesizing scientific achievements, man succeeded in analyzing more and more profoundly the mechanism which governs the evolution of various phenomena, and in forecasting their future trends; he was then able to take suitable actions, aimed at using for his benefit the development of processes which take place in nature and society.

The new scientific discoveries achieved in the twentieth century allowed man to subdue new phenomena of nature and society: nuclear energy, genetic structure, the extension of life, including active life, production of new chemical substances, direct investigation of the cosmos, development of automatic devices and data-processing equipment used for the optimal control of industrial processes, etc.

The qualitative changes which took place in the dynamics and the structure of contemporary society and of industrial production and technology appear more and more as results of the impact of modern science and technology on the reproduction process as a whole. The contemporary technical-scientific revolution represents a radical change in the development of productive forces; it marks profound changes in the whole social-economic structure. As opposed to the changes determined by the previous development of science and technology, the contemporary technical-scientific revolution speeds up the development of society characterized by permanent and accelerated changes in all domains of human activities". [M. Manescu, Economic Cybernetics, Tunbridge Wells, Kent TN. Abacus Press, 1980, pp. 1–2]

"Identification is conceived as a choice of a signal-meaning as relevant for a definite object or problem situation; the choice of a relevant sign would be at the same time the choice of a relevant (suitable etc.) object. The establishing of a relevant meaning is connected with concept formation. What is to be identified can be given in advance (e.g. by means of a pattern) or is to be found in the course of the process of cognition and solution. In solving identification tasks the human brain works with incomplete information, uses analogies, guesses, probability evaluation, removes uncertainty and superfluity. Accordingly, we consider discrimination of signs both relevant and irrelevant to be the starting point in the information analysis of identification and concept-formation processes. Understanding a sign of a cognised object and assigning meaning to the sign is a condition of identification. As mentioned above, identification in this sense means discrimination of relevant signs from irrelevant ones (independence on the choice of a viewpoint) and generalization of the relevant meaning. In information analysis we investigate further how much information is supplied by positive and negative instances, i.e. by knowledge that a definite stimulus is or is not an instance of the pursued concept. The choice of relevance is generally connected with adopting a subjective criterion in cognising an object" [R8.29, P.157].

"Perhaps the most fundamental concept of the universe is that of species of a population. A species is a set of individuals or objects each of which conforms to a common definition. The objects are not necessarily identical; indeed, in an important sense, no two objects can be identical because each occupies a different position in space-time. The definition identifies characteristics of an object that are similar enough to merit classification—as a species. This is taxonomy—the division of the universe into a partition of sets of objects, each of which constitutes a species. The definition is a kind of fence that includes all things in the species and excludes all things that are not. Every definition divides the universe into two parts: those things which conform to it and those things which do not. Taxonomy is part of our image of the universe, and it may be incorrect, in the sense that the species defined by the taxonomic definition are too heterogeneous to be a really interesting pattern. Thus, the medieval elements of earth, air, and water were too heterogeneous to be really useful in chemistry, and the chemical elements present a taxonomy that certainly increases the productive power of the human race over what could be done with the medieval elements" [R17.4b, p. 11].

"There is no doubt, of course, that information is associated with energy processes, that information interactions can take place only when specific correspondences between the energy of stimulation and energy of reaction exists, and that in order to transmit information a certain minimum quantity of energy is needed. Yet, to reduce information to energy processes alone is not justified. To transmit information is not possible without energy. But this still does not answer the question what is information and what is transmitted with the help of the specific energy processes" [R8.2, p. 28].

Washington, DC, USA Kofi Kissi Dompere

Prologue and Preface

The epistemic understanding of statics and dynamics in our knowledge–production system has not been clearly connected to information as an important property of matter and energy that define the system of varieties and changing varieties of being. Information is taken as something that resides in an amorphous space for use in all areas of socio-natural existence. Within this amorphous space arise epistemic conditions of statics and dynamics when time is conceptually acknowledged as a dimension of universal existence. The concepts of statics and dynamics find analytical meanings in the existence of varieties as established by information defined by a characteristic-signal disposition. Every construct of a dynamic system requires initialization of what it is and what it would be in terms of initial take-off identity and future-transformation identity. The point of take-off is called the initial conditions for the dynamic system and the efficient management of command and control decision-choice activities of the optimal trajectory under conditions of intentionality or non-intentionality.

Differences and Similarities of Ontological and Epistemological Dynamics

An important distinction must be established between *ontological dynamics* and *epistemological dynamics* in relation to four dimensions of matter-energy-information-time structure of the universal existence. The management of command-control decision-choice systems of ontological dynamics is under universal natural actions while the management of command-control decision-choice systems of epistemological dynamics is under cognitive actions.

The initial conditions are established to present the distribution of time-point identity distinctions in terms of differences and similarities of changing conditions of varieties and categorial varieties. The time-point conditions are in relation to the subsequent positional varieties of either a qualitative or quantitative dynamic behavior of the system of concern over its trajectory into the future identities. The collection of the initial conditions is a set of a general form including constant and variable parameters to specify the external information environment required for the specification of the systems descriptive initial positional variety. The specification of the initial set of conditions may be qualitative, quantitative or both to indicate the nature of initial identity of the variety or categorial varieties before the transformation dynamics. The initial identities are established by the initial characteristic-signal dispositions of the varieties or categorial varieties.

Some important analytical questions arise in the whole knowledge activities over the epistemological space. (1) What is the epistemic importance of specifying the initial conditions for qualitative-quantitative dynamic systems? (2) What is the meaning of the set of initial conditions in the general transformational dynamics? (3) Are the meanings of the sets of initial conditions the same for both ontological and epistemological transformations? (4) Similarly, are the meanings of the sets of the initial conditions of the ontological and epistemological dynamics different? If they are similar, what are the attributes of similarity and if they are different, what are the factors of difference? The theory of info-statics as developed in [R17.17] provides a framework to abstract some useful answers to these questions.

Explications of the Relevant Questions

Some of these questions need some explications in terms of the relevant exposition. The question of epistemic importance in specifying the set of initial conditions is to establish the nature of the initial time-position identity of any variety for comparative analyses of the distribution of time-position varieties of the same element over its trajectory or variety enveloping to assess the nature of the distribution of future time-position varieties in terms of their transformations for which identity valuations can be undertaken. In other words, the set of the initial conditions is a stock of information that accumulates with each time-position transformation. The set of the initial conditions is analytically used to specify the primary category (parent) from which successively derived categories are constructed as the offspring at each time point over the transformational trajectory. Under the specification of the set of the initial conditions as historic present, one can then, by comparative analytics, study the historic past by methodological reductionism as well as the historic future by methodological constructionism.

Every dynamic process has a beginning that may be characterized as the initial variety with its identity and information defined by a characteristic-signal disposition. The existence of the set of initial informational conditions holds for both ontological transformation dynamics and epistemological transformation dynamics. This existence is their common analytical link. The sets of ontological and epistemological conditions as revealed by their characteristic-signal dispositions are composed of subsets of negative and positive characteristics, and hence every variety has a negative and positive information that under appropriate entropic reflection may reveal negative and positive entropies over the epistemological space. It may be kept in mind that the existence of entropy makes has no meaning over the ontological space. The negativity and positivity of the information structure exist in duality, where every negative information has a corresponding positive information support and vice versa which may be projected into the structure of entropies. The negative and positive subsets of characteristics and information types are interchangeable in all conflicting transformation processes depending on the variety, its transformation state and the internal structure of the opposites. The negative-positive conflict relation is transformed to positive-negative conflict relation where negative assumes the attribute of positivity and positive is transformed into negative posture.

There are differences between the sets of ontological and epistemological initial conditions that may be examined. The first difference finds its expression in the idea that the set of ontological initial conditions for ontological dynamics of varieties is internally specified, while the set of epistemological initial conditions for the epistemological dynamics of varieties are externally and artificially specified by cognitive agents with cognitive intentionalities. In other words, there is an important difference in initialization responsibility. The second difference finds expressions in the sets of the initial conditions of ontological and epistemological transformations of varieties. At the level of qualitative motions of varieties, the set of ontological initial conditions for any ontological transformation may internally start at *zero*. This initial conditions of zero are the principle of ontological nothingness polarity under conditions of opposites, where from nothingness arose somethingness and into nothingness will everything vanish through a dualistic struggle[R17.19]. The condition of existence of nothingness-somethingess polarity is not opened to epistemological transformation, the activities of which fall into the command and control of cognitive agents. The set of epistemological initial conditions in qualitative transformation of any epistemological variety cannot be zero where zero conditions mean nothingness. In all epistemological transformations, this is the real economic principle of no-free-lunch condition or factor-indispensability condition. Simply stated, ontological transformation dynamics may be initialized at conditions of nothingness which is then taken as primary category of existence. The epistemological transformation dynamics cannot be initialized conditions of nothingness.

The existence of non-existence of nothingness-somethingness polarity is the ultimate fundamental question to understand the set of initial conditions of universal existence, the evolution, and expansion and transformation dynamics of the universe and the continual transformation of ontological varieties. It is here that science and religion intensely crash in search for the solution to the problem of the initial set of conditions of the universal existence and its dynamics. Whatever solution is designed to specify the set of the initial ontological conditions by whosoever, this solution must have an anchorage in the theory of info-statics in the sense that the initial conditions must provide the relevant set of info-stock for the initial universal existence as the primary category for all derived categories in terms of comparative identities and solution to the universal identification problem.

The Relational Nature Among the Theories of Info-statics, Info-Dynamics and General Transformation

The theory of info-statics provides all the analytical conditions of info-stocks for specifying the set of initial conditions for transformation of ontological and epistemological varieties and the corresponding info-stocks. The theory of info-statics establishers the definitional conditions for creating the required set of initial conditions for all dynamical systems and the corresponding-info-flows as explanations of which are presented as the theory of info-dynamics. In this sense, the theory of info-statics initializes the enveloping path of the dynamics of information production though the continual variety transformations. Within the epistemic structure, the *theory of info-statics* and the *theory of info-dynamics* combine to establish the *general theory of transformation* over both the ontological and epistemological spaces, where the theory of info-statics specifies the identification of the set of initial conditions and hence the set of initial varieties and categorial varieties as well as the identification of the sets of stock-flow conditions in the transformation process at each point of time of the changing characteristics of varieties and categorial varieties.

The Monograph

Given the theory of info-statics as providing the definitional framework and solution to the general identification problem in helping to specify the set of initial conditions of all dynamic systems, the core of the theory of info-dynamics is presented in explaining socio-natural transformations of varieties where such transformations generate info-flows with constant updating of the info-stocks as established by the conditions of the theory of info-statics. The conditions of info-statics provide a definition of information as composed of characteristic disposition for the recognition of the identity of each variety, thus providing a solution to the identification problem of the initial point of time and at each transformation time point in the enveloping process.

From the conditions of the info-statics in identity identification, the theory of info-dynamics provides philosophical and mathematical logical structure for the explanation of information production through variety transformations. The organic logical structure is constructed from the conditions of the principle of opposites which is composed of a system of infinite actual-potential polarities and corresponding infinite systems of negative-positive dualities with relational continua and unity. Every pole of an actual-potential polarity has a residing negative-positive duality in relational conflicts that generate energy and force from within the pole to execute the work of variety transformation in terms of internal destruction and creation under the principle of real cost-benefit conditions of actual-potential substitutability conditions.

The theory of info-dynamics that is developed in this monograph follows the logical framework that has been used as an epistemic tradition in the number of my works in the understanding of the knowledge-production process, decision-choice rationality, theory of knowing and actual-potential spaces as they relate to the possibility space, the probability space, and uncertainty-risk phenomena. The logical framework is expanded to include the role of real cost-benefit information in socio-natural transformations. The understanding is developed to bring into focus the logical relation of socio-natural transformational actions which generate qualitative and quantitative motions to bring about variety changes creating information flows to maintain levels of info-stocks as well as updating the info-stocks through info-flows under complex decision-choice processes. The info-stock-flow dynamics is a system of info-processes and processors through socio-natural transformations of varieties applicable to the understanding of the universal dynamics. The point here is that the system of change or transformation is simply a system of information production and info-flows without the destruction of info-stocks. The development of the monograph is to present a theoretical framework for the study of general dynamics in qualitative-quantitative time structure. This approach, therefore, unifies the understanding of all systems dynamics through information production. Given the goals and objectives of the monograph, the theory of info-dynamics is organized into six chapters.

The Chapter Summaries

Chapter 1 discusses the prelude conditions of the need for the development of the theory of info-dynamics. The chapter is used to reflect on the information theory in relation to variety transformations as seen in terms matter-energy structure where identities are presented as a combination of characteristic dispositions and signal disposition. The characteristic and signal dispositions are linked to the conditions of ontological and epistemological existences. The characteristic disposition holds the information contents while the signal disposition reveals the identity. Non-probability measures of information through the characteristic-signal dispositions are offered. The chapter is concluded with discussions on the relationships among variety, uncertainty, and information.

Chapter 2 is used to discuss some epistemic problems and questions in the general theory of information. The epistemic problems and questions in the general theory of information involve the relative constitution of the subject matters of the theories of info-statics and info-dynamics. Here, the concept the general uncertainty is discussed in relation to fuzziness and randomness in the outcomes of variety transformations. The concept of information production is linked to inter-categorial and intra-categorial conversions as defining the general variety transformation. The concepts of variety and category are defined and linked to the concepts of characteristic-signal disposition in the universal space to establish their structurally relevant relations for the development of the theory of info-dynamics which involves the interplay of thee dimensions of matter, energy and information in the universal existence.

Chapter 3 presents discussions of time as the fourth dimension of the universal existence to establish an analytical framework for the understanding of variety transformations in terms of qualitative and quantitative motions that show the complexities of interlinkages of matter-energy and information in terms of progress and retrogression. Different concepts of time are introduced. They involve the concepts of pure time with its corresponding time set, cost time with its corresponding time set, benefit time with its corresponding time set, cost-benefit time with its corresponding time set and transformation-decision time with its corresponding time set. These various concepts of time are defined, explicated and constructed to provide their analytic value for the development of the theory of the info-dynamics. The categorial indicator function is introduced to allow the time-point identification of varieties and categorial varieties in terms of change and no change, where the decision-choice of time point to transform over the epistemological space is formulated as a fuzzy decision-choice problem with the method of fuzzy optimization to abstract a solution which is then discussed around the information process.

Chapter 4 presents the introduction conceptual frame of the theory of info-dynamics. It begins with a general reflection on epistemic directions in the development of the theory. The reflections involve the establishment of the nature of the general transformation-decision systems over both the ontological and epistemological spaces. The structure of the ontological transformation-decision system is introduced as the primary category that performs the identity in the system of transformation processes in information production. The structure of the epistemological transformation-decision is then introduced as a derived category from the ontological structure.

The structure of the ontological transformation-decision system is constructed from a system of actual-potential polarities and then explained by the behavior of system of negative-positive dualities under the general principle of opposites, where the pathways of variety transformations producing information production is from the actual to the potential and from the potential to the actual under constructionism-reductionism methodological duality with relational continuum and unity. The structure of the epistemological transformation-decision system is also constructed from a system of derived actual-potential polarities and explained by a system of derived negative-positive dualities also under the principle of opposites, where the pathways of variety transformation go from potential to the possible, from the possible to the probable and from the probable to the actual and the to the potential under the actual-potential substitution principle. The explanation of the similarities and differences between these information generating processes are presented. The two structures must relate to the necessary and sufficient conditions for variety convertibility and transformability. The variety space is introduced and defined with explication and symbolic representation of the characteristic-signal disposition which is then linked to categorial conversion under pure and cost-benefit transformation-decision time. The chapter is concluded with a system of analytics for the theory of the socio-natural transformations and the corresponding introduction to info-algebra.

Chapter 5 presents extended discussions on the then algebraic introduction to the theory of info-dynamics. The concepts and the algebraic structure of the information process (info-process) and processors (info-processors) under socio-natural technological space are introduced with definitions and explications of the concepts of information process, information processor, information enveloping path and space with specified types of information processor. The concepts of forward (telescopic right) and backward (telescopic left) composition processors are algebraically presented and defined. Corresponding to them the concepts of information transformation transitional and information processor transitional matrixes are constructed to show the dynamics of varieties from the primary vector to the sequences of derived vector of varieties. The logic guarantees an infinite set of parent-offspring (ancestor-successor) processes. The matrixes are presented with explanations and linked to the concept of categorial conversion which establishes the necessary conditions of variety transformation and then to the concept of Philosophical Consciencism which establishes the sufficient conditions of variety transformations. The essential postulates emerging out of the theory are stated as seven laws of info-dynamics with discussions and comments. The chapter is concluded with discussions on info-stock, info-flow, and information disequilibrium under variety transformation, the intentionality of which finds expression in Philosophical Consciencism that guides the decision-choice actions over the epistemological space.

Chapter 6 relates the theory of info-dynamics to other knowledge areas where the theory is viewed as representing a framework of a unified theory of dynamics for all systems through information production. Here, there are presentations of info-dynamic reflections on thermodynamics, electrodynamics, energetics, and electromagnetism and Einsteinian fundamental equation in physics as an information representation of a claimed knowledge. The chapter is concluded with information stock-flow conditions and the roles of the family of ordinary language (FOL) and the family of abstract languages (FAL) play in representation and communication of information with discussions on the relative differences and similarities over the epistemological space.

Washington, DC, USA Kofi Kissi Dompere

Contents

Chapter 1
A Prelude to the Theory of Infodynamics: A Reflection and Critique of Information Theory

Some important works on research, learning and teaching have been done on the concept, phenomenon and measurement of information to produce a framework on how the phenomenon of information is traditionally viewed in philosophy, sciences and mathematics. This tradition has many approaches to the understanding of information. Each approach is an epistemic attempt over the epistemological space to conceptualize the phenomenon of information. The set of the results of each approach brings an illumination of the type of problem dealt with on the subject of information over the scientific-philosophical spaces. It is possible to view the phenomenon of information in either a static or a dynamic state. Both static and dynamic views may be seen in terms of either variety definition, identification system, information production, information transmission, information communication or all of the above. When information is viewed in terms of communication, even the scientist sees the totality of information phenomenon as devoiced from matter and connected to energy only in transportation between entities of source and destination. Here, the phenomenon of information is seen in term of quantitative disposition in the source-destination messaging process that is linked to objective probability, to the neglect of its qualitative disposition. Information acquires a mathematical definition, where the elucidation of the meaning and contents of information contained in the messaging systems in the transient processes come under the measurement and mathematical manipulation. When one says that information is a branch of mathematics, one implies that information is a language rather than seeing information as a phenomenon just as the phenomena of matter and energy the properties of which may be represented in a mathematical language. Languages are cognitive creations as representations of properties of diverse phenomena.

A question then arises as to what is been represented and what is been measured in the phenomenon of information. Is it the content that is been represented and measured or is it the meaningfulness of the content that is been represented and measured give any message? Similarly, when information is viewed in terms of communication at the philosophical space, even the philosopher sees the totality of

K.K. Dompere, *The Theory of Info-Dynamics: Rational Foundations of Information-Knowledge Dynamics*, Studies in Systems, Decision and Control 114, https://doi.org/10.1007/978-3-319-63853-9_1

information phenomenon as devoiced from matter and connected to energy only in transportation between the source and destination. Here, the phenomenon of information is seen in term of semantic qualitative disposition and with semantic definition that is linked to subjective probability and to the neglect of its quantitative disposition specified in terms of volume. The elucidation of the meaning and contents of information in the transient process comes under the true-false measurements and logical manipulation in exact space in terms of validity of the content which is not explicitly indicated. A question arises as to what is the true-false measurement intended to measure: Is it the content or the meaningfulness of the message?

When the phenomenon of information is viewed as concerning variety identification in the general space of knowing, then the totality of information phenomenon is fully connected to *what there is* and *what would be* and hence connected as a set of properties of matter and energy in quantitative and qualitative dispositions. In this respect, either the information transmission or communication in the source-destination process is about establishing distinction and commonness for knowing, learning, teaching and decision-choice activities under the principle of alternatives. Alternatives are meaningful within a set of varieties and similarities which in turn give analytical meaning to the concepts of indifference curves, isoquants, isothermal curves, and many others, all of which are designed to establish differences and similarities in qualitative and quantitative dispositions. The decision-choice process under the principle of alternative varieties is then connected to an information production to establish the essence of continual info-dynamics in the universal system without end. In this essence, knowledge and decision-choice activities are impossible without varieties.

1.1 The Framework of Reflections and Critique of the Tradition to Information Theory

The framework of reflections and critique of traditional theories on, and of information in this monograph is built on the notion of variety as the central epistemic tool in the concepts of identity, identity identification and identity distribution of elements and categorial elements. The framework demands that the concepts of variety, categorial identity and the distribution of categorial identities be philosophically defined and explicated for scientific usage. The framework for the critique further demands that the concept and phenomenon of information must be general with defined specificities that relate to the distribution of varieties and categorial varieties. The same framework must allow one to relate ontological identities to epistemological identities and how the relational structure of the two identities establishes the conditions of stock-flow dynamics of information and knowledge. The notions of ontological and epistemological identities must be clearly defined and distinguished in this framework of the concept and phenomenon of information. The concept and phenomenon of information must be independent of their representation, such as epistemic decoding of ontological encoding.

The available materials for epistemological analysis and decision are the signal dispositions which allow the establishment of distribution of epistemological varieties from which knowledge is claimed through their association to ontological characteristics and varieties. The set of problems of the concept and phenomenon of information must be separated from the set of problems of the ontological-epistemological transmissions and intra-epistemological communication of information. The first set of problems comes within the purview of the *material theory of information* to establish *what there is* in terms of content and meaning from the source. The second set of problems comes within the purview of the communication theory of information in terms of transmission of *what there is* from the source through a transmission mechanism to a destination. The general theory of information is composed of the material and communication theories about *what there is* and the transmission of *what there is* under the mechanism of source-destination interactive processes. *What there is* presents itself as a set of varieties and categorial varieties of ontological elements, the distribution of which presents elemental differences and similarities to establish varieties and identities for knowing and decisiding.

Given the concept and phenomenon of information, the content must be established for the possibility of designing a measurable unit for the quantity analysis of the amount of the content contained in the signal disposition. This measurement is relevant only over the epistemological space. It may be kept in mind that the signal disposition is a set of signals, where these signals relate to the elements of the characteristic disposition. These signals, for any variety, are encoded by the source agent and must be decoded by destination agents. The decoding process brings into the messaging mechanism two types of uncertainty of qualitative and quantitative dispositions expressed over the spaces of possibility and probability respectively over the epistemological space. The spaces of possibility and probability offer a framework to design a measure of volume of the content as well as the meaning of the content of acquaintance with the signal disposition depending on how the conceptions of qualitative and quantitative dispositions are analytically imposed to restrict the spaces. The inter-categorial varieties and the family of inter-categorial varieties and the corresponding acquaintances in the epistemic space are established by the distribution of qualitative dispositions. The design of any quantitative measure of the content of the acquaintance is a translation of the qualitative disposition into some form of quantitative disposition that is defined in the degrees of informing. There is a general theory of information that unites all conceptual forms of the phenomenon of information. This general theory of information is composed of the theory of info-statics and the theory of info-dynamics where the theory of info-statics initializes the understanding of information production in completely dynamic systems in the sets of macro-states and micro-states. The analytical structure of the *general theory of information* that is sought is presented as an epistemic geometry with resultant components in Fig. 1.1.

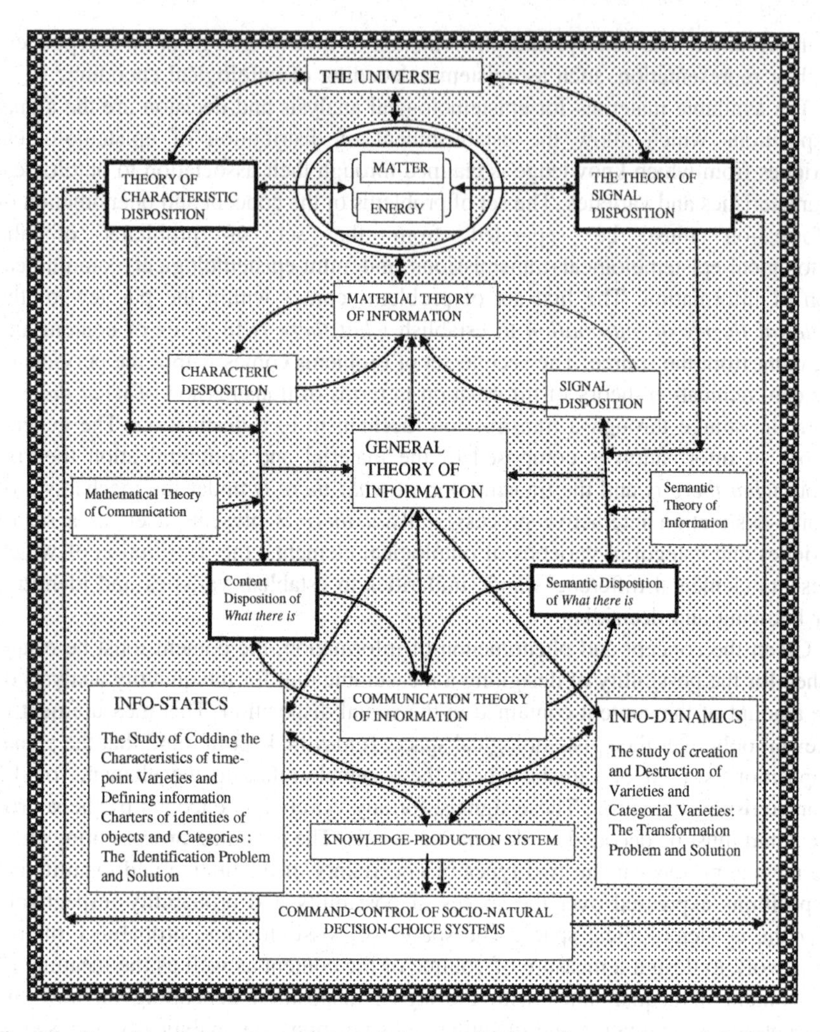

Fig. 1.1 An epistemic geometry of general theory of information compose of the theory of info-statics and the theory of info-dynamics

1.2 The Traditions on the Concept and Phenomenon of Information

There are few epistemic initiatives that must be discussed in an attempt to reflect on traditional and other views on the concept and phenomenon of information from the structure that have been provided in Fig. 1.1. They are qualitative and quantitative dispositions, materiality, and transmission over the ontological space, communications over the epistemological space and communications among elements in the ontological space and the elements of the epistemological space, and communications among elements of epistemological space and elements of ontological space. Here, a distinction is being made between the concepts of transmission and communication, where the concept of communication is contained in the concept of transmission. The understanding of the concept and phenomenon of information cannot be separated from their definitions and contents. The epistemic problems that are confronted are to identify the agency of the source and the acquaintances of the destination. The solution to the problems are important for examining the possible noises in the transmission mechanism and the available framework in constructing measures of the contents of information.

The transmission behavior over the ontological space is different from over the epistemological space. Over the epistemological space, the transmission behavior is defined in the context of noise system due to cognitive interactions. The behavior of such a noise system is not definable over the ontological space that holds identities of elements. Over the ontological space, the communications are multidimensional among ontological agents without noise from the source of ontological agents to the destination of ontological agents. The problem that arises in all transmission and communication is the nature of the codding from the source agent and the decoding by the destination agents. The decoding of the encoded message from the ontological source agent by the ontological destination agents may produce different contents through the medium of acquaintance and interpretations of the elements of signal disposition sent where different destination agents may have different interpretations of the same signal disposition. The differences are produced by the decoding mechanisms used by different ontological destination agents.

It is useful to consider the communications between ontological agents and epistemological agents. The encoding and decoding of messages are through the action on the distribution of the signal-dispositions given the characteristic disposition. The encoding action is carried on by an *encoding operator*. The decoding action is also done by a *decoding operator*. The signal dispositions are derivatives of the characteristic dispositions that define the distribution of the past-present varieties at any historic point. All the past-present history is always available at the ontological space that defines *what there is* in terms of varieties. In other words, *what there is*, is a collection of varieties. The identities of such ontological varieties are restricted to the epistemological space in terms of informing, knowing, analyzing, learning, teaching and using. The encoding of the signal dispositions from both the ontological source and epistemological source is perfect that reflects the nature of characteristic dispositions of the ontological and epistemological varieties. The decoding actions in the ontological space of both the ontological and epistemological encoding of the signal dispositions are perfect without noise. This is another way of saying that there is no uncertainty and risk in the ontological space.

The decoding actions through the decoding operators over the epistemological space of ontological signal dispositions are imperfect produced by cognitive and physical limitations on acquaintances that produce defective information of stochastic and fuzzy uncertainties which give rise to measures of possibility and probability as conditionality of degrees of exactness and riskiness respectively. These degrees of exactness and riskiness prevent the effectiveness of knowing in such a way that every claimed knowledge is under a fuzzy-stochastic conditionality [R4.13]. The encoding and decoding actions over the epistemological space of the epistemological encoding-decoding process and among the epistemological agents from the epistemological signal dispositions come with increasing imperfections of defective and deceptive structures of volume limitations, fuzzy limitations, disinformation and misinformation. These attributes produce combinations of fuzzy-stochastic uncertainties and riskiness in abstracted knowledge and knowledge search by the destination agents from the epistemological signal dispositions as transmitted by the epistemological source agents. The volume and fuzzy limitations shape the qualitative and quantitative directions of the variety transformations and the information production over the epistemological space through the decision-choice system and the corresponding outcomes.

A number of questions arise under the conditions of defectiveness and deceptiveness of decoding the elements of the signal disposition by the epistemological destination agents from the messages of the elements as sent by the ontological source agents as well as from the messaging elements from the epistemological destination agents to the epistemological destination agents. How should one measure the quantitative disposition relative to qualitative disposition? The question as to what is being measured must be answered with some conditions of degree of exactness. This is the content problem of all concepts. Here, the unit of measurement is not important, the only requirement in the epistemic discourse is consistency of the unit and the measurement. How does this measure affect the nature of informing, awareness and understanding of differential situations with neutrality of time as past, present and future are connected? In other words, how meaningful is the element contained in the message to the destination agent and is the decoding isomorphic to the encoding? Here, arises the semantic disposition and analysis of meaning and possible measure of degree of meaningfulness of the message to the destination agent and the concern of the source agent about the degree of accuracy in decoding the message by the destination agent.

The search for meaningfulness contained in a message and the degree of implied meaningfulness are in fact the foundational problem for the development of semantic theory of information. In this respect, the semantic theory of information is in reference to qualitative disposition of messaging. Similarly, the question of how should one measure the quantitative disposition relative to qualitative disposition gave rise to the development of mathematical analysis and mathematical theory of information. Here, one is not concerned with meaningfulness and the degree of meaningfulness of a message. One is concerned with the fullness of the volume of the message. The solution to the problems of measurement of information are not different from the solution to the measurement problems of other phenomena such as forms of energy, matter, efficiency and many others. The general communication process of signal disposition is presented in Fig. 1.2.

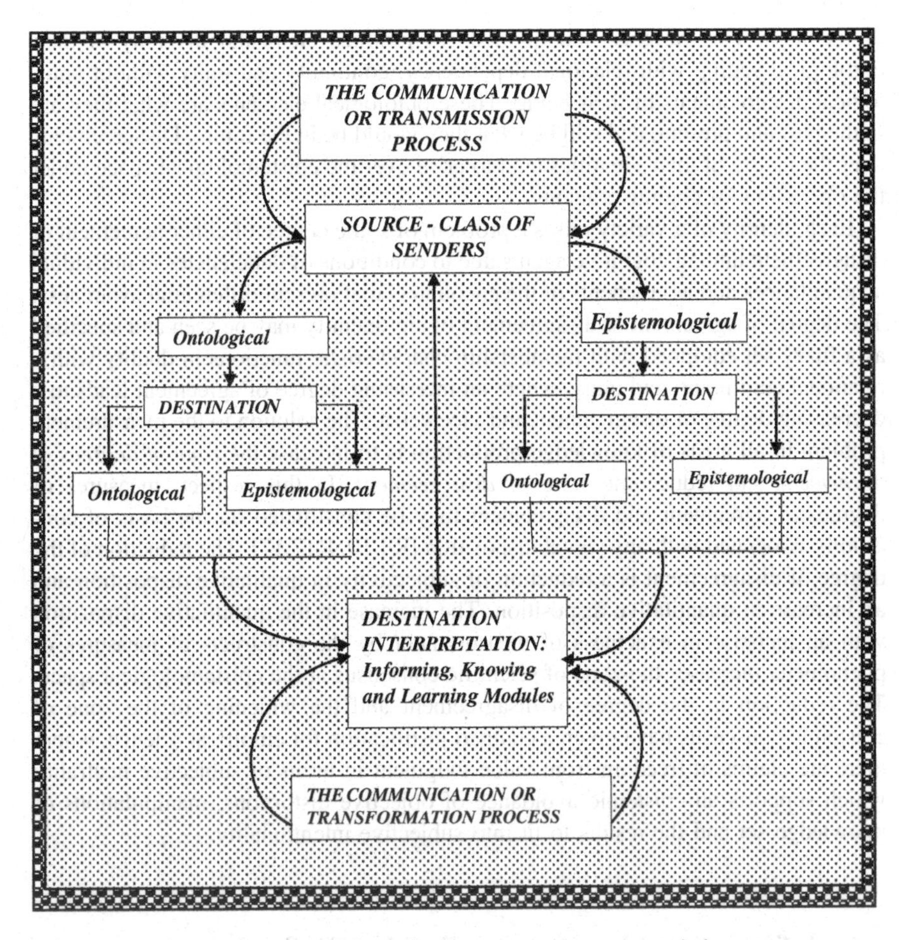

Fig. 1.2 An epistemic geometry of the path of communication or transmission process

1.3 Measurements and Their Relevance in Science and Knowing

Any socio-natural phenomenon is composed of qualitative and quantitative dispositions in continuum and unity. The qualitative and qualitative dispositions present themselves in differential combinations in each socio-natural elements to create a distribution through the activities of their internal energies. The nature of the distribution of the deferential quality-quantity combinations imposes a second order distribution of socio-natural identities of varieties and categorial varieties in dividedness. The varieties and dividedness produce elemental diversity at their ontological setting with relational dependency, continuity and unity without ontological disagreements of their being. These ontological elements are what they are in the past and present, and will be what they would be in the future. Disagreements over the nature of their identities tend to arise over the epistemological space under the operations of cognitive agents.

Each identity of a variety raises a problem of some degree of either agreement or disagreement among cognitive agents due to conditions of deficiencies in individual acquaintances and assessments of the encounters that present differential awareness. The degree of agreement or disagreement of any identity may be seen and specified as a fuzzy set equipped with membership characteristic functions in duality with a relational continuum and unity, where an increasing degree of agreement leads to a decreasing degree of disagreement and vice versa. The solution to the measurement problem in all areas of knowledge production is to increase the *zone of degrees of agreement* and reduce the *zone of disagreement*. In this respect, measures of degrees of agreement and disagreement may be constructed on the basis of conditions associated with any signal disposition [R4.10, R17.17]. The reduction of the degree of disagreement is a reduction of subjectivity in qualitative disposition and an increase in quantitative disposition. The increase in the quantitative disposition with an increase in exactness of measurement by the constructs of measures of phenomena increases the zone of scientific consensus at the epistemological space. The reduction in the degree of disagreement and the widening of the zone of epistemic consensus have given a claimed justification for empirical science, where there are claims of freedom from value judgment in terms of subjective neutrality which supports an epistemic avoidance of objective distortions, manipulations of knowledge set and the results to fit into subjective intentionality.

1.4 Information-Measurement Problem Defined

The measurement problem in knowledge production over the epistemological space may be seen in terms of characteristic sets that present a general distribution of identities in the process of acquaintance for informing, learning and knowing of varieties and categorial varieties. The universe is teeming with an infinite system of phenomena Φ with a generic element $\phi \in \Phi$ that generates an infinite set of universal complexities associated with elements of universal object space Ω with a generic element $\omega \in \Omega$. The universal phenomenon and object space have an infinitely corresponding universal characteristic set $\mathbb{X}$ with a generic element $x \in \mathbb{X}$. The universal characteristic set exist as a duality made up of a positive characteristic sub-set $\mathbb{X}^{P}$ and a negative characteristic sub-set, $\mathbb{X}^{N}$ such that $\mathbb{X} = \left(\mathbb{X}^{N} \cup \mathbb{X}^{P}\right)$ and that the intersection may or may not be empty, $\left(\mathbb{X}^{N} \cap \mathbb{X}^{P}\right) \lesseqgtr \varnothing$ to allow for a relational continuum and unity. For each phenomenon $\phi_i \in \Phi$ and objective $\omega_i \in \Omega$ there is a corresponding characteristic sub-set $\left(\mathbb{X}_{\phi_i} \subset \mathbb{X}\right)$ for $\forall i \in \mathbb{I}^{\infty}$ which is an infinite index set for the elements except when indicated. The corresponding characteristic sub-set presents the identity of a variety $v_i \in \mathbb{V}$, which specifies the space of the actual and possible universal varieties. The universal characteristic set is a negative-positive duality of subsets of elements. Each $\mathbb{X}_{\phi_i}$ characteristic sub-set is composed of negative characteristic sub-set $\mathbb{X}^{\mathrm{N}}_{\phi_i}$ and positive characteristic sub-sets $\mathbb{X}^{\mathrm{P}}_{\phi_i}$ such that $\mathbb{X}_{\phi_i} = \left(\mathbb{X}^{\mathrm{N}}_{\phi_i} \cup \mathbb{X}^{\mathrm{N}}_{\phi_i}\right)$ with the condition that $\mathbb{X}^{\mathrm{N}} = \bigcup_{\phi_i \in \Phi} \mathbb{X}^{\mathrm{N}}_{\phi_i}$, $\mathbb{X}^{\mathrm{P}} = \bigcup_{\phi_i \in \Phi} \mathbb{X}^{\mathrm{P}}_{\phi_i}$ and $\mathbb{X} = \left(\bigcup_{\phi_i \in \Phi} \mathbb{X}^{\mathrm{N}}_{\phi_i}\right) \cup \left(\bigcup_{\phi_i \in \Phi} \mathbb{X}^{\mathrm{P}}_{\phi_i}\right)$. By the principle of relational continuum and unity of duality and polarity, it is not the case that $\left(\bigcup_{\phi_i \in \Phi} \mathbb{X}^{\mathrm{N}}_{\phi_i}\right) \cap \left(\bigcup_{\phi_i \in \Phi} \mathbb{X}^{\mathrm{P}}_{\phi_i}\right) = \varnothing$. It is analytically useful to keep in mind that every negative has a positive support and vice versa, the nature of which is constrained by the general existing conditions. The positive and the negative elements in the transformation dynamics may not only be inter-changeable but inter-transformable [R17.15, R17.16, R4.13] in the general conceptual framework of cost-benefit conditions of elements and time where every cost has a benefit support and vice versa. In other words, the negative-positive duality is not different from the cost benefit duality defined in the fuzzy space under the *Asantrofi-anoma rationality* where every decision-choice action comes with inseparable cost-benefit conditions in the decision-choice transformation process [R17.15].

The universal characteristic set $\mathbb{X}$ also exists in a dualistic form as sub-sets of qualitative (Q = qualitative) and quantitative (R = quantitative) dispositions which are shared with the negative (N) and positive (P) characteristic sub-sets. In this respect, the negative universal characteristic sub-set may be partitioned into a qualitative negative characteristic sub-set $\mathbb{X}^{\mathrm{NQ}}$ and a quantitative negative characteristic sub-set $\mathbb{X}^{\mathrm{NR}}$ such that the term $\mathbb{X}^{\mathrm{N}} = \left(\mathbb{X}^{\mathrm{NQ}} \cup \mathbb{X}^{\mathrm{NR}}\right)$ defines negative qualitative and quantitative dispositions. Similarly, there is a partition of the qualitative characteristic set in the form of a qualitative positive characteristic sub-set $\mathbb{X}^{\mathrm{PQ}}$ and a qualitative negative characteristic sub-set $\mathbb{X}^{\mathrm{PR}}$ such that the term $\mathbb{X}^{\mathrm{P}} = \left(\mathbb{X}^{\mathrm{PQ}} \cup \mathbb{X}^{\mathrm{PR}}\right)$ defines positive qualitative and quantitative dispositions. The conditions of qualitative and quantitative dispositions are transferred to the characteristic sets of specific phenomenon $\phi_i \in \Phi$ and object $\omega_i \in \Omega$.

The transferring process induces a partition over $\mathbb{X}_{\phi_i} = \left(\mathbb{X}^{\mathrm{Q}}_{\phi_i} \cup \mathbb{X}^{\mathrm{R}}_{\phi_i}\right)$. The negative characteristic sub-set for any phenomenon $\phi_i \in \Phi$ may also be specified as $\mathbb{X}^{\mathrm{N}}_{\phi_i} = \left(\mathbb{X}^{\mathrm{NQ}}_{\phi_i} \cup \mathbb{X}^{\mathrm{NR}}_{\phi_i}\right)$ with the corresponding positive characteristic sub-set specified as $\mathbb{X}^{\mathrm{P}}_{\phi_i} = \left(\mathbb{X}^{\mathrm{PQ}}_{\phi_i} \cup \mathbb{X}^{\mathrm{PR}}_{\phi_i}\right)$. The assumption of conditions of interaction whether crisp or fuzzy will depend on the epistemic situation and the corresponding problem. Let us keep in mind that all these sets and subsets are defined in terms of characteristic dispositions in the ontological space. The categorial interactive conditions may be presented with Venn diagram as in Fig. 1.3. The general structure is under the principle of opposites in duality and polarity with relational continua and unity that lends itself to the acceptance of fuzzy partitioning and identity definition.

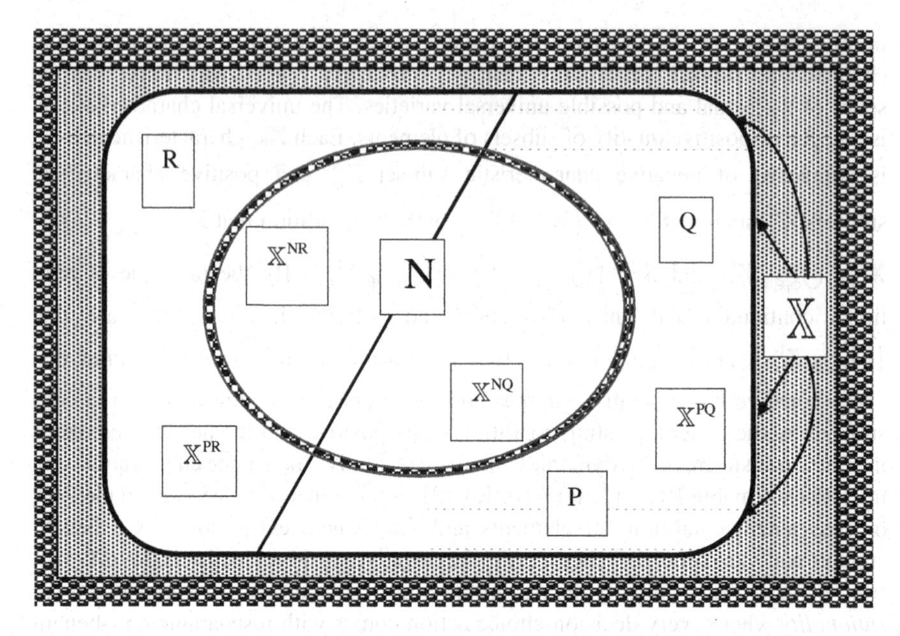

Fig. 1.3 Venn diagram for the division of the universal characteristic set $\mathbb{X}$: P = Positive, N = Negative, Q = Qualitative Disposition, R = Quantitative Disposition, $\mathbb{X}^{NQ}$ = Negative Qualitative Disposition, $\mathbb{X}^{NR}$ = Negative Quantitative Disposition, $\mathbb{X}^{PQ}$ = Positive Qualitative Disposition, $\mathbb{X}^{PR}$ = Positive Quantitative Disposition

The nature of the identity may be abstracted from the diagram shown in Fig. 1.4. Any variety is revealed by a specific identity through its characteristic disposition. The characteristic set is divided into negative and positive characteristic subsets with the imposition of qualitative and quantitative dispositions. The relative qualitative-quantitative disposition supported by the relative negative-positive characteristic set defines the identity of the variety, the knowledge of which is encoded in the signal disposition and sent across communication or transmission channels and made possible to all ontological objects with acquaintances over the epistemological space.

From the structure of Fig. 1.4, the phenomenon with its object $(\phi \in \Phi, \omega \in \Omega)$ is said to be qualitatively defined if $\#\left(\mathbb{X}_{\phi}^{NR} \cup \mathbb{X}_{\phi}^{PR}\right) < \#\left(\mathbb{X}_{\phi}^{PQ} \cup \mathbb{X}_{\phi}^{NQ}\right)$. It is said to be quantitatively defined if $\#\left(\mathbb{X}_{\phi}^{NR} \cup \mathbb{X}_{\phi}^{PR}\right) > \#\left(\mathbb{X}_{\phi}^{PQ} \cup \mathbb{X}_{\phi}^{NQ}\right)$. It is said to be indeterminate and subjectively determined if $\#\left(\mathbb{X}_{\phi}^{NR} \cup \mathbb{X}_{\phi}^{PR}\right) = \#\left(\mathbb{X}_{\phi}^{PQ} \cup \mathbb{X}_{\phi}^{NQ}\right)$. The solution process to the measurement problem over the epistemological space is a decision-choice process to increase $\mathbb{X}_{\phi}^{R}$ and reduce $\mathbb{X}_{\phi}^{Q}$ for all phenomena $\phi \in \Phi$ and the corresponding objects $\omega \in \Omega$.

The solution to the measurement problem in the quality-quantity duality with relational continuum and unity involves the process of finding either an absolute or

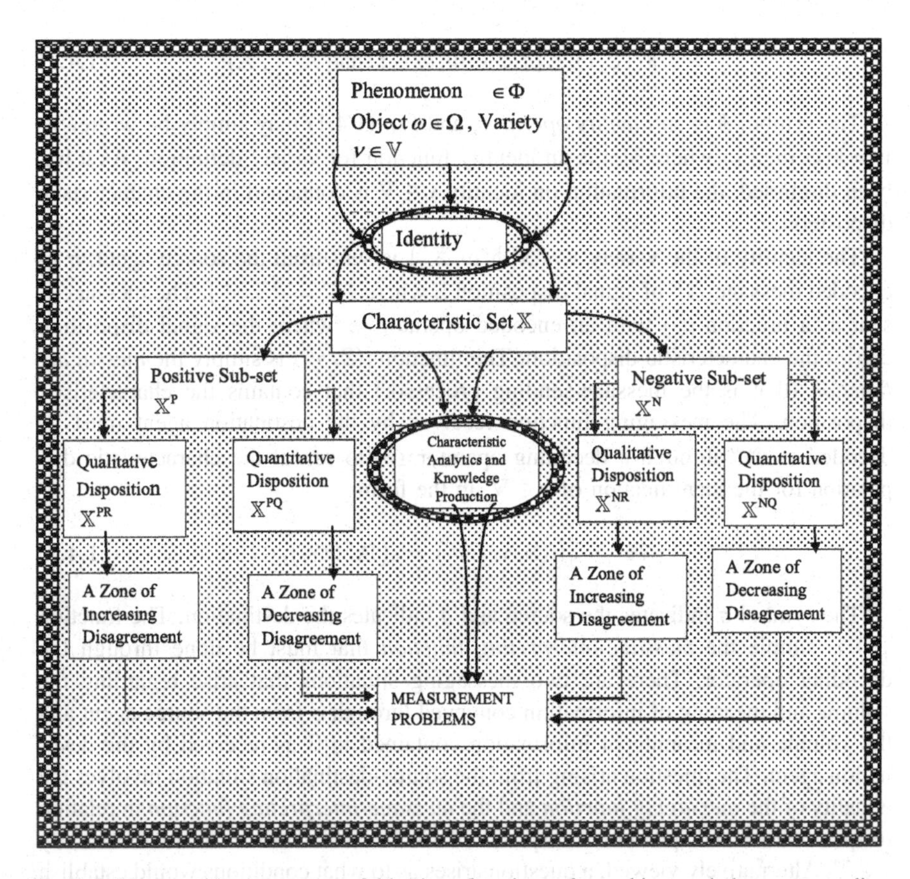

Fig. 1.4 The cognitive geometry of identity of variety of an object and the corresponding phenomenon

ordinal measure with some unit of measures, depending on whether the characteristic set of the variety is dominated by the qualitative disposition or the quantitative disposition. The cardinal and ordinal measures may appear as an absolute or a relative value. The absolute measure and its unit are easily favored because of the ease with which they can be associated with quantitative dispositions and the reduction of disagreements in applications. The measurements and the corresponding units may be seen in terms of distribution of identities of varieties. The units of measure, thus define a framework to establish varieties of measures associated with varieties of phenomena and universal objects. It is useful to keep in mind, at this point, that every variety $\nu \in \mathbb{V}$ exists and is defined by a phenomenon $\phi \in \Phi$ with a corresponding object $\omega \in \Omega$ to establish objects, where the establishments of objects, phenomena and varieties are through the characteristics $\mathbb{X}$. In other words, any variety $\nu \in \mathbb{V}$ may be written as:

$$v = \mathfrak{I}((\phi, \omega, \mathbb{X}_\phi) | \phi \in \Phi, \omega \in \Omega \text{ and } \mathbb{X}_\phi \subset \mathbb{X}) \quad (1.4.1)$$

The symbol $\mathfrak{I}$ is an *identity operator*, and Eq. (1.4.1) may simply be written as $v = \mathfrak{I}(\phi, \omega, \mathbb{X}_\phi)$ where $\mathfrak{I}(\cdot)$ is an identity function for variety definition. As it has been discussed, the information on identity $v \in \mathbb{V}$ is contained in the characteristic disposition.

The identity of the variety $v = \mathfrak{I}(\phi, \omega, \mathbb{X}_\phi)$ is sent from the source agent, σ as encoded message $\mathfrak{M}^\sigma$ of signal disposition $\mathbb{S}$ where $\mathfrak{M}^\sigma = \mathfrak{f}(\mathbb{S}, \mathscr{E})$. The message is sent as a function $\mathfrak{f}(\cdot)$ with an encoded operator $\mathscr{E}$ where the signal disposition carries the characteristic disposition. The function $\mathfrak{f}(\mathbb{S}, \mathscr{E})$ is simply the *messaging function* (that is the message-sending processor) that contains the characteristic disposition. The messaging function received by the destination agent must be decoded as $\mathfrak{M}^\delta$ through a decoding operator $\mathscr{D}$ to reveal the characteristic disposition for the identification of $v \in \mathbb{V}$, in the form

$$\mathfrak{M}^\delta = \mathfrak{g}(\mathfrak{f}(\mathbb{S}, \mathscr{E})\mathscr{D}) = \mathfrak{g}(\mathfrak{M}^\sigma, \mathscr{D}), \quad (1.4.2)$$

The symbol σ indicates the source and δ indicates the destination. The function $\mathfrak{g}(\cdot)$ is simple the *message-interpreting function* that must be done through the decoding operator. There are two messaging situations that arise regarding the volume and meaning of information conveyed through $\mathbb{S}$ from the source agent and the volume and meaning of information contained in $\mathbb{S}$ as received by the destination agent. The question arising from the source-destination situation involves the equality of the *messaging function* and the *message-interpreting function* in relation to quantitative and qualitative dispositions of the characteristic set of the variety $v \in \mathbb{V}$. Alternatively viewed, a question arises as to what conditions would establish the equality between the message from the source and message received at the destination $\mathfrak{M}^\sigma = \mathfrak{M}^\delta$? In other words, is the message sent from the source correctly interpreted by the destination and hence the interpreted message is isomorphic to the message sent relative to the contents?

In order to speak of the volume (quantity) and meaning (quality) of information sent and received in some scientific sense, it is useful to raise some questions, the answers of which will bring some analytical clarity. Are there differences among the concept of information, the phenomenon of information, the information transmitted from the source agent and the information received by the destination agent? Are there differences among the measures of the concept of information, the information transmitted from the source and information received at the destination? If there are differences, what are they and how do they arise and how can we account for them? What are the relational structures among information, possibility, probability, the associated epistemic uncertainties and corresponding measures? Here, two organic types of possibilistic and probabilistic uncertainties arise. Are these uncertainties the result of the nature of information surrounding phenomena and hence defined by information or do these uncertainties define the meaning and nature of the concept and phenomenon of information? Are these uncertainties source-transmission derived or destination-interpretation derived? At some point in the discussions, the transmission process shall be distinguished from the communication process when ontological information is distinguished from the epistemological information.

1.5 Uncertainty, Variety and Information

When the concept and phenomenon of information are used to establish the meaning and existence of uncertainty, information becomes the primary category of knowing while uncertainty becomes the derived category of knowing which gives rise to the concept and phenomenon of risk as information-derived. In this case, the distribution of universal varieties and the path of knowing correspond to the distribution of the characteristic-signal dispositions where information is claimed as an essential part of matter just as energy, and that any definitional structure of information must fulfill this universal role. In this conceptual system, information is also a property of energy through matter. Similarly, when the concept and phenomenon of uncertainty are used to define the existence of information, then uncertainty becomes the primary category of knowing while information becomes the derived category of knowing which gives rise to the conceptual system, where all things are defined and known. In this alternative view, the distribution of universal varieties and the path of knowing correspond to the distribution of uncertainties as conceived and defined.

A problem arises as to whether uncertainty is ontological or epistemological? If uncertainty is ontological, then it must be shown that it is an organic property of matter and energy at the level of static states of elements. In the epistemic frame that is being projected, uncertainty is epistemological and arises through cognitive limitations associated in informing, knowing, deciding and choosing of varieties. The concepts of information and uncertainty are composed of quality and quantity that are reflected in linguistic measures as linguistic variables for inter-categorial and intra-categorial comparisons and ordering of varieties where quantitative and qualitative dispositions exist in a relationally inseparable continuum and unity. The cognitive activities of distinguishing and categorizing varieties for inter-categorial and intra-categorial comparison and ordering require information at the static level. The development of the required theory for the nature and behavior of these static properties of categories belongs to the subject matter of *info-statics*. The general structure of the theory of info-statics has been presented in a companion monograph on definitional foundation of information [R17.17].

1.5.1 Uncertainty and Categories of Measures of Concepts and Phenomenon

In the general knowledge production regarding concepts and measurements, one can conceive of three measurement problems of exactness, inexactness and the combination of the two and the instrumentation that may be imposed. The demand of the development of theories of measurement is over the epistemological space where cognitive agents operate over the space of cognitive actions with epistemic limitations. The exact, inexact and exact-inexact measures are numerical and non-numerical in structure to deal with the comparability problems of inter-categorial and intra-categorial ranking of varieties in understanding and utilization. Since every elemental variety exists in quality-quantity duality, and hence in an exact-inexact duality, it is analytically useful to speak of measurements in terms of degrees of exactness which place them in a fuzzy space of reasoning. The relational structure of quantity-quality and exactness-inexactness dualities under the principles of relational continuum and unity allows an epistemic examination of information and its communication in terms of *volume* and *meaning* within the source-destination duality, where every source has a destination support and vice versa. It is useful to work with an analytical principle that the volume and meaning of message from the source agent are information exact and complete even though it may contain deceptive components at some level of intentionality as have been explained in a monograph entitled the theory of Philosophical Consciencism [R17.16].

The message contains encoded information defined by characteristic-signal disposition which constitutes the *primary category* at the source. The characteristic disposition presents the contents which is carried by the signal disposition. The destination decodes the message into an interpreted message which becomes the *derived category* at the destination. The concepts of primary and derived categories of informing, knowing and learning in the decision-choice systemicity and complexity must be clearly understood if one is to understand the theories of info-statics and info-dynamics and how these theories relate to categorial existence of matter and energy. A newly defined conceptual space arises when the primary and derived categories are related to the measurement problem over the epistemological space.

At the level of the sources, the measurement problem faced by any source agent is not the degree of exactness but the capacity of transmission of the required encoded information. This is the *capacity limitation principle* of message transmission, the measurement of which will depend on the *information length* and *encoding efficiency*. At the level of destinations, the measurement problem faced by a destination agent is not that of transmission capacity but the degree of exactness of volume and meaning of information decoded from the source message needed for processing into knowledge as an input into the decision-choice system. This is *the interpretive capacity* of the received message, the measurement of which will depend on the *degree of deception* and *decoding efficiency*. The specific meaning and applications of some of these terminologies will vary from messaging of ontological information from the ontological space to the epistemological space and the intra-epistemological messaging of epistemological information. The degrees of deception induced by disinformation and misinformation are applicable over the intra-epistemological messaging system and not over the ontological-to-epistemological messaging system.

The problem of capacity limitation on encoded information does not arise for ontological source over both the ontological and epistemological spaces. The ontological characteristic-signal dispositions are what they are and will be what they would be. The characteristic dispositions are encoded in the signal dispositions from the source and to be decoded at the destination. For example, the distribution of the ontological varieties is established by the distribution of the ontological characteristic dispositions. The ontological characteristic dispositions are encoded by nature into the ontological signal dispositions for a corresponding distribution of signal dispositions that holds the distribution of the identities of the varieties from the source. Each identity resides in a source-destination duality over the ontological space. The encoded information of each variety is sent as a general message over the ontological space to all destination recipients who must interpret the message into information through a process of decoding. For cognitive agents, this process takes place over the epistemological space. It is these source-destination and encoding-decoding mechanisms that some elements of ontological varieties are discovered through a process to form a social knowledge. Ontological varieties reside in ontological encoding. Epistemological varieties reside in epistemological decoding. The problem of exactness does not arise for ontological destinations except for the destinations of cognitive agents who operate over the epistemological space to decode the ontologically encoded signal dispositions to find the corresponding characteristic dispositions and the associated varieties in the process of seeking to acquire some degree of independence of existence from the ontological space through an acquaintance and awareness.

1.5.2 Reflections on the Measurement of Information

The measurement of information and its unit of measure are relevant only over the epistemological space, where the rise of qualitative disagreements establishes some domain of subjective phenomenon such that order-ranking inconsistency may arise among cognitive agents. To the source agent the need for measurement is to determine the relevant and appropriate channel capacity for the transmission of the signal disposition that carries the characteristic disposition of a variety or a categorial variety. The required channel capacity will depend on the size of the signal disposition which is equipped with an encoding operator that must be decoded for understanding and meaning. The decoding of the signal disposition reveals the underlying structure of the characteristic disposition and the identity of the associated variety. The information contained in the decoded message at the destination is said to be perfectly transmitted if it is completely isomorphic to the information contained in the encoded message from the source. In this way $\mathfrak{M}^{\sigma} = \mathfrak{M}^{\delta}$ and the information from the source becomes the information at the destination and the communication is perfectly synchronized. The message sent and the message received are said to be information isomorphic. The conditions for the information isomorphism, among other things, will depend on the relational structure of encoding operator $\mathscr{E}$ and decoding operator $\mathscr{D}$ given the transmission mechanism. An example may be found in all languages such that there exists a *translation function operator* $\mathcal{k}$ that maps the decoding operator onto the encoding operator with $\mathcal{k}(\mathscr{D}) = \mathscr{E}$ and $\mathscr{D} = \mathcal{k}^{-1}(\mathscr{E})$. An examples of transmission mechanism in the case of languages are voice and writing. Knowledge production requires some degree of information isomorphism between ontological encoding and epistemological decoding.

A number of questions tend to arise concerning the communication activities of cognitive agents over the epistemological space. Given a channel capacity, what is the size of an optimal signal disposition that may be sent through both a noiseless or a noisy channel to a destination and what is the associated risk of transmission error? Given a signal disposition of defined variety, what is either an optimal noiseless or noisy channel capacity required to transmit it with minimal chance of transmission error? What does it mean to speak of transmission error or risk in both cases? Are the transmission errors and risks defined by the concept and measure of probability and does probability define the conditions of uncertainty and hence can one use probability to define the concept of uncertainty and vice versa? Is the measurement problem of information communicated and received the same as the measurement problem of information in general whether communicated or not? These questions relate to the question whether probability and its measure are the appropriate epistemic vehicles for defining and measuring information over the epistemological space and can these vehicles be extended to deal with the information conditions over the ontological space? The answers to these questions are part of the theory of info-statics which are treated in the [R17.17].

The concept and phenomenon of general information in the epistemic process have been divided into ontological information and epistemological information with distinguishing characteristics that place identities of universal elements and define them in existential varieties. The nature of their transmission are different where information transmission is applied to ontological information and information communication is applied to messaging systems by cognitive agents over the epistemological space. The communication among epistemological agents takes as an intra-epistemological messaging process. The study of transmission of ontological information to the epistemological space leads the understanding of information processing the results of which lead to an ontological knowledge of *what there is* in varieties. The communication of information through the messaging system is mostly applied to epistemological information. There are two types which are social information and natural information. The natural epistemological information is a direct derivative from the ontological information through the principle of acquaintance with the ontological signal disposition. The ontological information is the production of relations among ontological elements. The social information is the production of social relations of all forms which include misinformation and disinformation that generate propaganda in social decision-choice relational systems. In this respect, the methods and techniques for the study of intra-epistemological messaging system require some modifications from the methods and techniques for the study of ontological-to-epistemological messaging system.

Chapter 2
Some Fundamental Epistemic Problems and Questions in General Information Theory

There are a set of epistemic problems and corresponding questions that may be reflected on in order to provide a useful environment for the understanding of the general conditions of static theory of information which is referred to in this monograph as the *theory of info-statics*. The theory of info-statics defines the initial epistemic conditions for the construct of *the theory of info-dynamics*. In other words, it initializes the dynamics of information as an inseparable property of matter and energy. The epistemic question that arises is what are the subject matter of info-statics and the subject matter of info-dynamics. What are the differences and similarities among the subject matters of info-statics and info-dynamics? Similarly, one may want to know the differences in methodology and the corresponding techniques and methods that the methodological framework generates for the development of the theories and the meaning analytics.

2.1 The Subject Matters of Info-statics and Info-dynamics

The theory of info-statics is a branch in the theory of knowledge concerned with definition of the concept of information that will establish the information phenomenon and its content, measurement and communication. Furthermore, the theory of info-statics must deal with the nature of qualitative and quantitative dispositions of information and how they relate to universal objects to define their identities, meaning, distinction, similarities and differences at any point of time. Under these conditions, the theory of info-statics must provide the required framework and conditions to form categories of varieties at any time point. In other words, it must include a category theory that defines the conditions of its formation and analysis through categorial analytics. The category formation may be constructed through the methods and techniques of fuzzy decomposition of universal existence of matter and energy. In this way, the theory over the epistemological space must deal with the information properties that make it possible for

K.K. Dompere, *The Theory of Info-Dynamics: Rational Foundations of Information-Knowledge Dynamics*, Studies in Systems, Decision and Control 114, https://doi.org/10.1007/978-3-319-63853-9_2

comparative analytics, order analytics and decision-choice analytics. Over the ontological space, the theory of info-statics must deal with conditions of existence and its observability over the epistemological space in time. Furthermore, the theory of info-statics must provide an understanding of the relational structure among matter, energy and information under similarities and differences which establish identities of varieties and categorial varieties of the collection of universal phenomena $\phi \in \Phi$ and the collection of the corresponding elements $\omega \in \Phi$ at a defined time position. The theory of info-statics must also establish a sequential structure at any time point where the theory presents conditions in a manner that allows the identification of the primary categories on the basis of which other categories of knowing may be sequentially derived to define the enveloping of qualitative and quantitative dispositions with neutrality of time. In other words, the theory info-statics must establish the initial necessary and sufficient conditions for one to know if a change of a variety has occurred with different time identities.

In all these respects, the epistemic problem that is confronted by the development of any analytical structure of the theory of info-statics is to establish an appropriate sequential relational structure among information, uncertainty and corresponding measures that are epistemologically possible. In the previous volume [R17.17], sequential definitions of the concepts of uncertainty, data, fact and knowledge were provided. The concept of information was defined as a characteristic-signal disposition and was taken as the primary category from which the concepts of uncertainty, data, fact and knowledge are defined as logical derivatives at any given time point. The characteristic-signal dispositions and their distribution fix identities of varieties and their distributions into categories at any point of time. The characteristic-signal disposition is a static concept on the basis of which each time-point variety is identified and distinguished. The existence of identities of varieties and categorial varieties and their distributions as presented by information in terms of characteristic-signal dispositions makes knowledge construction and the art and science of knowing over the epistemological space possible.

The *concept of uncertainty* was defined as a condition in a situation which is characterized by either *incomplete information*, *vague information* or both surrounding the epistemological identity of a variety under the principle of acquaintance over the epistemological space. The concepts of incompleteness and vagueness are only, thus, applicable over the epistemological space and not over the ontological space. The incomplete information is related to quantitative dispositions of signal dispositions which in turn are related to random variables, probability measures and stochastic environment surrounding conditions of the identities of observed varieties within the time-point decision-choice systems. The conditions of vague information is related to qualitative dispositions of signal dispositions which in turn are related to fuzzy variables and fuzzy measures or linguistic variables and linguistic measures, and possibilistic environment surrounding conditions of the identities of observed varieties within the time-point decision-choice system. The simultaneous existence of the incomplete and vague information is related to quantitative-qualitative dispositions and find expressions in either fuzzy-random or random-fuzzy variables with either fuzzy-probability measures or probabilistic-fuzzy measures that define a special information environment for identities of observed varieties within the time-point of decision-choice system and all epistemological activities. In this epistemic frame, decision-choice activities are impossible without knowledge; knowledge is impossible without information; information is impossible without an acquaintance and acquaintances are meaningless without varieties.

The question that arises is how do the measures of uncertainties when they are constructed, provide levels of epistemic understanding of the concept and phenomenon of information, the establishment of identities of elements in terms of varieties and the distribution of varieties under the principle of nominalism with the development of knowing over the epistemological space. To examine the question of relationality of measures of uncertainties to the understanding of information phenomenon, it is useful to keep in mind that different forms of uncertainty are epistemological and not ontological. The forms of uncertainty, vocabulary, relational meanings of things are impossible without a clearly defining structure of the concept and phenomenon of information. Information is the fuel of meaning and knowing, knowledge is the derivative of knowing as well as the fuel of learning while logic and learning are the fuels of teaching and communication under cognitive limitations.

Uncertainty is a derived phenomenon, where the associated object is not directly connected to matter and energy. The concept of uncertainty is a derived category from information which constitutes the primary category of knowing. Without knowing, learning and teaching under decision-choice activities over the epistemological space, the concept of uncertainty is undefined and empty. Like any phenomenon, uncertainty is composed of qualitative and quantitative dispositions with neutrality of time. How is uncertainty related to the concepts of possible and probable and hence possibility and probability? How may these linguistic values be translated into quantitatively defined values and be associated with information? In other words, what does it mean to say something is either possible or probable or both? In the world of knowing, the words, possible and probable must be given clear meanings and mutually distinguished in varieties. At the scientific level, however, they must be explicated by the way of axioms or measurements or both, where the clarity must find expressions in the concepts of scientific importance through definition and explication. In knowledge production and mathematical discourse, what are the differences and similarities of the concepts associated with probable and possible in other to establish analytical variety?

The words in a language structure acquire same meanings in ordinary communications among members of the same linguistic group. The meanings of words are associated with concepts and projected onto phenomena and then related to objects for understanding and decision-choice action. The differences and similarities of two words of interest must be established in both ordinary and scientific communications in the language of their existence. In ordinary communications, both probable and possible are associated with likelihood which is related to some degrees of cognitive unsureness in outcomes and understanding of that which is communicated. The likelihood is related to decisions of phenomena of outcomes of socio-natural processes as well as decisions on the meaningfulness of what is communicated. In other words, the concept of likelihood presents conditions of similarity of possible and probable which are then projected onto the spaces of possibility and probability respectively in the decision-choice spaces regarding outcomes and linguistic meanings within the constructed epistemological information structure over the epistemological space.

All these epistemic activities and associated decision-choice actions are possible with some implicitly or explicitly defined concept and phenomenon of information. The likelihood and unsureness find expressions in the phenomenon of uncertainty relative to information defined as characteristic-signal disposition to establish differences and similarities that are specified in the outcome space of socio-natural processes within the ontological space. The similarities and differences must also be defined in the meaning sub-space of communication space over the epistemological space regarding both ontological information and epistemological information as discussed in the theory of info-statics [R17.17]. It is the phenomenon of information that links meaningfulness of language and decision-choice actions together to establish the continuity of past, present and future relative to socio-natural transformations. It is also the phenomenon of information that allows an automatic and non-automatic control systems to be defined, established and managed.

There is another element of similarity between possible and probable from which a difference may be established. This element of similarity is the *power* that flows from *energy*. Possible and possibility find expressions in a capacity to hold power to effect outcomes in the transformation processes in actual-potential socio-natural polarities. Probable and probability find expressions in ability to use power to affect and effect outcomes in the transformation processes in the actual-potential polarities through the behaviors of the residing dualities. The former is the *power capacity* and assessed in terms of the degree of *capacity power sufficiency* for all phenomena $\phi \in \Phi$ and the collection of the corresponding elements $\omega \in \Omega$ at any fixed point in time, where $(\phi, \omega) \in (\Phi \otimes \Omega)$. The degree of *capacity power efficiency* is expressed in terms of possibilistic uncertainty. The degree of power-capacity sufficiency is the *power capacity principle* in transformations and outcomes where every outcome is a variety among the set of potential varieties defined by different information structures. The assessment of possibilistic uncertainty resided in the spaces of vagueness and subjectivity of qualitative dispositions relative to quantitative dispositions. The distribution of the degree of power capacity sufficiency provides possibility distribution for possibilistic uncertainty in the potential space. The *capacity power sufficiency* establishes the *necessary conditions* for transformation of varieties while the *capacity power efficiency* establishes the *sufficient conditions* for transformation of varieties. The necessary conditions relate to necessity of transformation in terms of categorial conversion [R17.15] and the sufficient conditions relate to freedom in transformation in terms of Philosophical Consciencism [R17.16].

The latter is power-utilization ability and assessed in terms of degree of power-resource usage efficiency for affecting and effecting the identity of phenomenon, $\phi \in \Phi$ and the corresponding object $\omega \in \Omega$ given the existence of power capacity. The degree of power-utilization efficiency is expressed in terms of probabilistic uncertainty. The degree of power-usage efficiency is the *power-usage efficiency principle* in transformation and outcomes, where every outcome is a variety among the set of possible varieties defined by information structures. The assessments of the probabilistic uncertainty resides in the space of volume, subjectivity, objectivity and approximation of qualitative-quantitative dispositions in relational continuum and unity. The distribution of the power-usage efficiencies provides the underlying conditions for the probability and distribution of probabilistic uncertainties in terms of outcomes of varieties in the transformation process. Let us keep in mind that the relational structure of matter, energy and information constitutes the foundations in the creative-destructive process of varieties where information performs two important roles of input of the management of commands and controls of the complex decision-choice systems to bring about the transformation of one variety to another as well as the role of input to know whether a change has taken place or not. The transformation process requires the combined structure of energy *capacity-sufficiency principle* and energy *usage-efficiency principle* in the spaces of qualitative and quantitative dispositions which must be related to necessity and freedom of sovereignty in decision-choice systems.

In this epistemic analysis, possibilistic uncertainty reflects defective information structure in assessing the capacity of a process to affect or effect an actualization of a potential variety of a phenomenon and its corresponding object $(\phi, \omega) \in (\Phi \otimes \Omega)$. Possibilistic uncertainty reflects defective information structure in assessing the ability of a process to exercise the capacity to either affect or effect an actualization of a variety of a phenomenon and its corresponding object $(\phi, \omega) \in (\Phi \otimes \Omega)$. It is through this relational understanding of power-capacity and power-ability that one finds meaning in concepts of possible and impossible outcomes of varieties and zero probability in probabilistic concept of reasoning as associated with the general cognitive decision-choice outcomes in the sense that zero-power capacity provides no ability to either affect or effect the transformative behavior of variety of interest. In this respect, the distribution of degrees of power-capacity may be associated with the probability distribution over an outcome of a variety. When one considers the past-present-future time structure in relation to static and dynamic conditions of information, one must analytically distinguish the concept of ontological outcome from that of epistemological outcome where such a distinction also involves the concept of information. The relational structure of possibility and probability has been explained and extensively discussed in [R4.7, R4.10].

2.2 Information, Inter-categorial and Intra-categorial Conversions

It has been pointed out that any socio-natural transformation involves necessary conditions in terms of *necessity* that relates to energy-capacity sufficiency defining the *capacity principle* in the works of transformations, and sufficient conditions in terms of *freedom* that relates to the energy-usage efficiency defining the *utility principle* in the works of transformations. The conditions of necessity is established by categorial conversion while the conditions of freedom is established by Philosophical Consciencism internal constituencies of objects. On all cognitive planes, the processes of informing, knowing, teaching and learning are about continual interactions among information, decision and results, where information is a gross input into socio-natural processes about epistemic distinctions and similarities of socio-natural varieties. The cognitive establishment of similarities and differences allows an epistemic creation of socio-natural categories with methodological nominalism and language for intra-categorial and inter-categorial comparisons of varieties and the corresponding identities. The inter-categorial differences relate to differences in qualitative dispositions, and provide an information foundation for epistemic studies of qualitative dynamics in terms of qualitative-time processes for stationary and non-stationary creation and distinctions among qualitative varieties for any given quantitative disposition. This is the quality-time phenomenon, the study of which leads to the development of the *theory of info-dynamics* at the level of qualitative disposition of which this monograph is concerned.

The study of info-dynamics at the level of qualitative variety is an inter-categorial conversion or inter-categorial info-dynamics, where the quantitative disposition is held constant. The intra-categorial differences relate to differences in quantitative dispositions, and provide an information foundation for epistemic studies of quantitative dynamics in terms of quantitative-time processes for stationary and non-stationary creation and distinctions among quantitative varieties for any given qualitative disposition. This is the quantity-time phenomenon, the study of which leads to the development of the *theory of info-dynamics* at the level of quantitative disposition of which this monograph is also concerned. The study of info-dynamics at the level of quantitative variety is intra-categorial conversion or intra-categorial info-dynamics where quantitative varieties change holding a qualitative disposition constant. Both qualitative disposition and quantitative disposition may change simultaneously to increase the complexity in the epistemic analysis and the construct of the theory of info-dynamics in the sense that there are simultaneity in the quantity-quality information stock-flow dynamics. There are plenty of examples in all the three cases in socio-natural transformations.

It may be analytically emphasized that the theory of info-dynamics is intimately connected to the theory of socio-natural transformations of varieties that presents an explanation of information stock-flow processes. The theory of socio-national transformation is divided into two sub-theories of categorial conversion of varieties and Philosophical Consciencism that presents the internal decision choice dynamics. The theory of categorial conversion presents an explanatory system of the necessary conditions for transformation. This is the categorial-conversion problem and its solution. The theory of Philosophical Consciencism presents an explanatory system of the sufficient conditions of transformation. This is the philosophical Consciencism problem and its solution. The general theory of transformation is the combined explanatory systems on necessary and sufficient conditions for the quantitative-qualitative dynamics of varieties and categorial varieties. The categorial-conversion problem with its solution and the Philosophical-Consciencism problem with its solution constitute the general transformation problem and its solution. The intra-categorial differences relate to quantitative dispositions of varieties, holding the qualitative disposition constant, at any point of time thus defining static conditions of quantitative identities. In other words, intra-categorial difference are established within given qualitative categorial varieties. This is the quantity-time phenomenon which includes the development of the theories of space-time dynamics. The inter-categorial differences and similarities of varieties provide informational foundations for epistemic studies of qualitative statics and dynamics in terms of qualitative-time processes at stationary and non-stationary states of qualitative dispositions.

In the process of understanding socio-natural events, it is useful to study them in terms of info-statics and info-dynamics in relation to varieties as seen in terms of categorial elements. The identity of every variety is defined by its conditions of info-statics. The knowing transformation of this identity to any other identity is defined by its conditions of info-dynamics. In this epistemic framework, mathematical and non-mathematical theories of category provide useful tools to study the general concept and phenomenon of information in time and over time. The theory of info-statics allows one to develop comparative analytics and form categories on the principles of difference and similarity of characteristics of varieties. Categorial, group and set analytics are impossible without varieties the characteristic-signal dispositions of which offer the conditions to perform their formations in time and transformations over time.

At the level of statics and over the epistemological space, the theory of info-statics may be divided into a theory of category formation and a theory of language including syntactic, signaling and other forms of representation such as spoken and written at any point of time. Language is nothing more than encoding-decoding process of the characteristics of varieties. The theory of category formation establishes categorial varieties and identities on the basis of acquaintance and distribution of characteristic-signal dispositions where the characteristic-signal dispositions provide defining attributes of the phenomenon and content of information. The theory of language establishes vocabularies of varieties and communication of varieties under the principle of rules of combination in time and over time. Language is not possible without real varieties. Vocabulary is nothing but a collection of linguistic varieties and similarities as provided by the collection of characteristic-signal dispositions. Decision and choice acquire no meaning under conditions of commonness and sameness of objects.

At the level of dynamics, the development of the theory of info-dynamics is initialized by the static conditions of information as established by the theory of info-statics. Given the static conditions, the theory of info-dynamics is composed of two sub-theories of the theory of *qualitative dynamics* and the *theory of quantitative dynamics*. There is a third theory which is the combination of the two that presents complexity in every system. This is the theory of simultaneous movements of qualitative and quantitative motions of the same object overtime. The theory of qualitative dynamics is about inter-categorial transformation of varieties where varieties shed off their initial characteristics in a fundamental way through internal transformation and move from one category to another. The internal process of an object to bring about a change in variety through a change in identity is called *categorial conversion* to create a continual information flow in the flow disequilibrium process of information. The theory for studying this information dynamics of varieties is called the *theory of categorial conversion* that maintains *stock-flow disequilibrium* dynamics over an infinite time.

The theory of qualitative dynamics is in a class of all cognitive frames developed on the principle of constructionism-reductionism duality in the quality-quantity space involved in transformations of varieties from within. This class includes theories of socio-economic development and all areas of engineering such as medical, biological, electrical, mechanical and many others. The theory of qualitative dynamics, thus, defines the epistemic foundations for the development of a general theory of unity of engineering sciences. With respect to this, engineering is viewed as a process of creation and destruction of varieties and continual generation of information flow and updating micro-macro information stock through the internal acts of differentiations and counter-differentiations in a non-stopping mode within the universal system. Qualitative movements or transformations are governed by qualitative equations of motion. Quantitative movements or transformations are governed by quantitative equations of motion. The paths of these equations of motion must always be initialized through the initial characteristic-signal dispositions called the initial time-point varieties that may be taken as the primary categories in order to know the starting points that provide relative time-point derived varieties.

The theory of quantitative dynamics is about intra-categorial changes of quantitative varieties from within holding qualitative dispositions of the same varieties constant within a category and defining a new quantitative disposition. Here, it is useful to observe that the theory of quantitative dynamics is also a class of all cognitive frames developed on the principle of constructionism-reductionism duality in quantity-time space involving changes of quantitative disposition of the same qualitative varieties under conditions of time. Any quantitative variety is defined by a quantitative characteristic-signal disposition that provides a specific information on identity such as *position and time*. The explanatory system of the behavior of this class of position-time varieties includes theories on time-space phenomena with constant qualitative disposition. With regard to quantitative disposition, each point in the space is a quantitative variety of the same qualitative variety. Changes in any quantitative disposition is governed by a quantitative equation of motion that may be self-induced or externally induced. The theory of quantitative dynamics, thus, defines the epistemic foundations for the development of a general theory of unity of sciences of physical motions in all linear and nonlinear spaces. With respect to this, science of physical motion is viewed as a process of creation and destruction of *time-position varieties* and continual generation of quantitative information flow and updating the quantitative information stock through position differentiation and counter-differentiation in a non-stopping mode in the universal system under conditions of time.

Quantitative motion is simply a position transformation where such a transformation provides an information flow in relation to position-time varieties of the same qualitative disposition of identity. The time-position varieties are distinguished by and information set containing the initial distance conditions, velocity, acceleration, distance, momentum and other important characteristics including conditions of the universe, galaxy, and celestial bodies. With respect to this epistemic frame, two quantitative varieties are said to be the same if the position of quantitative variety is unchanging with respect to a given reference point of a body and hence it is said to have the same quantitative identity or position-indifference. In a more general frame, a quantitative motion is a concept that applies to objects, bodies, and particles of matter, to radiation, radiation fields and radiation particles, and to space, its curvature and space-time at the constancy of qualitative disposition.

Much scientific works have been done on behavior of elements, positioned varieties in the quantity-time space for the study of quantitative static states and quantitative equations of motion. In the intra-categorial conversion, the subject matter is on the distributional time positions of characteristic-signal dispositions of a given phenomenon and the corresponding object of the form $(\phi, \omega) \in (\Phi \otimes \Omega)$ which characterizes the identity of the variety $v \in \mathbb{V}$ with information $(\mathbb{X}, \mathbb{S})_{\mathbb{C}} \in (\mathbb{X} \otimes \mathbb{S})$. The term $(\mathbb{X}, \mathbb{S})_{\mathbb{C}}$ is the information on varieties in category $\mathbb{C}$ which may also be written as $(\mathbb{X}_{\mathbb{C}}, \mathbb{S}_{\mathbb{C}})$. The subject matter of Info-statics deals with the initial conditions of quantitative and qualitative varieties $v \in \mathbb{V}$, phenomena and corresponding objects, $(\phi, \omega) \in (\Phi \otimes \Omega)$ and characteristic-signal dispositions $(\mathbb{X} \otimes \mathbb{S})$ and categories $(\mathbb{X}, \mathbb{S})_{\mathbb{C}}$. The theory of info-statics, therefore, is the study of initial conditions of varieties. It thus involves the definitional conditions and structures of varieties at a given time point. These initial conditions allow one to establish identity and categorial identity and their corresponding distributions. It is impossible to observe qualitative and quantitative transformations of varieties without their initial identities. These initial identities are defined by characteristic-signal dispositions, where quantitative-time varieties and qualitative-time varieties are studied as info-dynamics.

The subject matter of Info-dynamics deals with changing conditions of varieties from their initial conditions of quantitative and qualitative varieties $v \in \mathbb{V}$, phenomena and corresponding objects, $(\phi, \omega) \in (\Phi \otimes \Omega)$ and characteristic-signal dispositions $(\mathbb{X} \otimes \mathbb{S})$ and categories $(\mathbb{X}, \mathbb{S})_{\mathbb{C}}$. The theory of info-dynamics, therefore, is the study of continual transformation of conditions of varieties and information process. It thus involves the definition and redefinition of conditions and structures of varieties over continual time points. These changing conditions allow one to establish transformational identities and categorial identities and their corresponding transformational distributions. The theory of info-dynamics is thus the study of family of families of functional systems as changing information relations with the passage of time. It is the study of intra-categorial and inter-categorial conversions of varieties in terms of destruction of existing varieties and the creation of new varieties leading to a continual production of information flows and info-stock-flow disequilibrium.

The previous monograph was devoted to the static conditions of information under the title The Theory of info-statics: Conceptual Foundations of information and knowledge [R17.17]. In the current monograph, however, an attention is devoted to the development of the theory of info-dynamics. The combined theories of info-statics and info-dynamics will provide us with an epistemic environment to the understanding of not only unified sciences, unified engineering sciences and unified social sciences but the understanding of information as a property of matter. Before turning to the task, it may be stated that when one examines the phenomenon of information, from a general perspective in relation to matter and energy, one arrives at an epistemic form that a complete theory of information is not part of applied mathematics. Mathematics is a representation and study of information. It is in this respect that mathematics is viewed as an abstract language with neutrality of ordinary linguistic structures. Similarly, information is not part of applied physics, chemistry or other natural sciences or applied social sciences such as economics. These areas are varieties in the knowledge system and find distinctions and meaning from information.

Different socio-natural sciences are the study of specific and different forms of information. Information is not part of theoretical and applied engineering. Theoretical and applied engineering is the study and use of specific and different forms of knowledge derived from information. Information is not part of communication theory. Communication theory is the study of messaging systems of information over the epistemological space, the contents of which are established by characteristic-signal distributions that establish varieties and categorial varieties. Communication of information is an activity and contained in the information transmission as a general activity. This distinction between communication and transmission will be made explicit in this monograph. Information is not part of theory of languages. The theory of languages is about the theory of information representation that makes it possible for the development of the theories of communication and transmission in diverse forms. All these theories are about the study of different forms of varieties that are established by forms of characteristic-signal disposition of varieties. The existence of socio-natural varieties makes a search for *what there is* and *what would* be meaningful and possible.

It is analytically fruitful to be mindful of the notion that none of these theories are constructible without information and hence one can speak of the theory of matter-energy-information system which encompasses all theories of specific areas of informing and knowing through the study of the distribution of characteristic-signal dispositions as foundations of information systems. Information systems are systems of varieties and categorial varieties with corresponding identities and categorial identities. In terms of categories of existence, the category formation from awareness, knowing, teaching and learning, the theory of information is the primary category while all actual and potential theories over the epistemological space are derivatives within the constructionism-reductionism duality with relational continuum and unity. The point of special emphasis is that one cannot use any of these theories from different specific area of knowing about varieties to establish the general concept, content and phenomenon of the *universality of information*. This is because information is a matter-energy dependent property of universal existence. Nothing is knowable or teachable over the ontological space or epistemic space without the variety-identity concept of information. Similarly, all languages are impossible without the variety-identity concept of information.

Over the epistemological space, that cognitive agents function, the concept of information is made intelligible in the source-destination duality by the signal dispositions, where the contents of each signal disposition is defined by the corresponding characteristic disposition. The distribution of characteristic dispositions presents different varieties of matter and energy that can be known through the distribution of signal-dispositions over the epistemological space. The signal disposition of each variety is an encoded representation of the characteristic disposition from the source and must be decoded by the destination agent. The production of matter-energy varieties is continuous and hence generates static and dynamic conditions of the distribution of characteristic-signal dispositions that define information to reveal the distributions of varieties $v \in \mathbb{V}$, phenomena and corresponding objects, $(\phi, \omega) \in (\Phi \otimes \Omega)$, characteristic-signal dispositions $(\mathbb{X} \otimes \mathbb{S})$ and categories $(\mathbb{X}, \mathbb{S})_{\mathbb{C}} = (\mathbb{X}_{\mathbb{C}}, \mathbb{S}_{\mathbb{C}})$. In this epistemic frame, the initialized time-point value of information for ontological objects is to create conditions that make it possible to know and act in relation to subjectively defined time-point goal-objective set. The study of every dynamic system demands that one knows the sets of the initial conditions, where the sets of the initial conditions present the initial time-point varieties of the objects with a corresponding initial identity in such a way that the elements in the distribution of the time-point identities of the same objects with corresponding distributions of time-point varieties can be distinguished for differences and similarities over the relevant time period where the identities of the time-point varieties are made up of velocity, acceleration, position and time in quality-quantity domain. The study of variety-transformation is simply the study of qualitative and quantitative time-point enveloping of universal objects where such enveloping reveals the information on the dynamic behaviors of objects.

2.3 Information, Category Formation and Categorial Conversion

The theory of info-statics relates to the establishment of initial conditions of transformation of varieties. This initial conditions find expressions in the general definitional characteristics of information [R17.17]. It, thus, allows the determination of micro-macro information stocks at different time points. The transformation conditions of a variety are related to changes in the time-point identity and a creation of new time-point characteristic-signal dispositions that define a variety-production to generate a flow of information and the corresponding changes in the levels of information stocks. The flow of information is a time-phenomenon and governed by equations of motion in the quality-quantity space. The analysis of transformation and continual changes of varieties must be seen in terms of the dynamics of information stock-flow conditions. The stock-flow conditions of information must be connected to matter-energy relationality in the universal system of things as seen in terms of varieties and categorial varieties. Every transformation of a variety is a production of information flow and an updating of information stock in the universal system. The theory of transformation of varieties is the theory of information production and corresponding flows in terms of changes of the corresponding characteristic-signal dispositions.

To examine transformations of varieties, identities must be established as well as categories must be formed. These require the development of a *theory of info-statics* which is composed of the *information definitional foundations of variety and category formation* of varieties. The category formation depends on the use of a properly defined concept of information in order to establish a set of varieties $v \in \mathbb{V}$, where the concept of information as characteristic-signal disposition allows a creation of differences and similarities. From the information of differences and similarities, categories of varieties are formed as categorial varieties in the form $(\mathbb{C}_v | v \in \mathbb{V})$ where the category $\mathbb{C}_v$ is formed at a fixed $v \in \mathbb{V}$. The categories may be seen as identity-indifference curves that establish iso-variety sets. In this way, the universe Ω is decomposed by $v \in \mathbb{V}$ into a space of varieties with corresponding identities. This is the variety decomposition principle of universal existence. Under an epistemic action, the variety decomposition of the universal object set Ω may be linked to the fuzzy decomposition through the fix-level (α-level) set where Ω is viewed as a fuzzy space of varieties which is equipped with a membership characteristic function $\mu_{\Omega}(v)$. This will be discussed when the concept of entropy is taken up.

The development of the theory of category formation is only possible under the theory info-statics. The usage combination of the theories of info-statics and category formation provides the initial conditions to develop the understanding of transformation of varieties in a broad general sense of qualitative and quantitative changes. The explanatory and prescriptive understanding of the process of transformation of varieties is undertaking by the *theory of categorial conversion.* The assessments of validity of a transformed variety are done by comparative analytics of the characteristic-signal dispositions that are associated with a given variety $v \in \mathbb{V}$ at different time points. The sizes of the spaces of varieties, categories and information stock at each time point depend on time and transformative successes as defined by equations of motion in the quality-quantity space. Let a note be taken that at any time point the size of the stock of information is greater than the sizes of the space of varieties and categorial varieties. This is easily justified by the principle of disappearance of the old and the emergence of the new. The old variety disappears but its information structure does not disappear and is retained as part of the information stock. The new variety brings in a new information which is then added to the information stock to increase its level. The categorial conversion is the creative-destruction of varieties as well as defining the information flow in a flow disequilibrium process. The categorial formation supported by categorial conversion finds expressions in information *stock-flow disequilibrium processes.*

To deal with the phenomenon of information stock-flow disequilibrium, it is useful to provide workable definitions and explications of the concepts of category formation and categorial conversion and how they provide entry points to the development of the theory of info-dynamics. Category formation is the development of static conditions that make it possible to distinguish varieties and place them in categories. Categorial conversion is a time-dependent process that relates to same element and its changing behavior over time. The development of the theories of category formation, categorial conversion and Philosophical Consciencism, and the understanding of their inter-dependent relational structure are the foundations of the theories of info-statics and info-dynamics. Let us keep in mind the definition of information from the theory of info-statics developed in [R17.17], where the concept and phenomenon of information at any time point are fixed by the characteristic-signal disposition of the form $(\mathbb{X}, \mathbb{S})_{\mathbb{C}} \in (\mathbb{X} \otimes \mathbb{S})$. The $\mathbb{X}$ is the characteristic disposition and $\mathbb{S}$ is the signal disposition. The distribution of identities of varieties $v \in \mathbb{V}$, is fixed, at any given time point, by the distribution of characteristic-signal dispositions. In terms of epistemic foundations of knowing, learning, teaching and communication, all areas of knowledge production is about the study of epistemic–signal dispositions to identify, understand, know, communicate and teach varieties.

Definition 2.3.1 (*Category*)
Given the universal object set, $\omega \in \Omega$, the corresponding set of phenomena $\phi \in \Phi$, the characteristic set, $x \in \mathbb{X}$ and the signal set, $s \in \mathbb{S}$, let $(\approx)$ be an *identicality* relation defined over $\mathbb{X}$ and $\mathbb{S}$ for each pair $(\omega, \phi) \in (\Omega \otimes \Phi)$ then the group, $\mathbb{C}_k$, is said to be a category if and only if there exists $(\omega_1, \omega_2, \ldots, \omega_k) \in \Omega$ such that $(\mathbb{X}_1 \approx \mathbb{X}_2 \approx \cdots \approx \cdots \mathbb{X}_k) \in \mathbb{X}$, and $(\mathbb{S}_1 \approx \mathbb{S}_2 \approx \cdots \approx \cdots \mathbb{S}_k) \in \mathbb{S}$ then $(\omega_1, \omega_2, \ldots, \omega_k) \in \mathbb{C}_k$, with a corresponding $(\phi_1, \phi_2, \ldots, \phi_k) \in \mathbb{C}_k$. Alternatively, if there exists $(\omega_1, \omega_2, \ldots, \omega_k) \in \mathbb{C}_k$ then the corresponding characteristic sets are identical in the sense that, $(\mathbb{X}_1 \approx \mathbb{X}_2 \approx \cdots \approx \cdots \mathbb{X}_k) \in \mathbb{X}$ and so also the signal sets in the form $(\mathbb{S}_1 \approx \mathbb{S}_2 \approx \cdots \approx \cdots \mathbb{S}_k) \in \mathbb{S}$ with identical information of the form $\mathbb{Z}_k = (\mathbb{X}_k \otimes \mathbb{S}_k) = \{z_j = (x_j, s_j) | j \in \mathbb{J}_k \subset \mathbb{J}^{\infty}, k \in \mathbb{L}^{\infty}\}$ that defines the

characteristic-signal disposition in the quality-quantity-time space such that $((\mathbb{X}_1, \mathbb{S}_1) \approx (\mathbb{X}_2, \mathbb{S}_2) \approx \cdots \approx \cdots (\mathbb{X}_k, \mathbb{S}_k)) \in (\mathbb{X} \otimes \mathbb{S})$.

Note
The definition of a category may be used to explicitly establish definitions of the concepts of variety and categorial variety which will help in examining the general changing conditions of information and their relationships to the subject matter of info-dynamics. The definition of category implicitly contains the concept of variety and categorial variety, where every object with the corresponding phenomenon $(\omega_k, \phi_k) \in (\Omega \otimes \Phi)$ is described by a corresponding characteristic-signal disposition $\mathbb{Z}_k = (\mathbb{X}_k \otimes \mathbb{S}_k)$. The definition of the concept of variety must establish conditions of difference and similarity that allow an element $(\omega_k, \phi_k) \in (\Omega \otimes \Phi)$ to be examined against an element $(\omega_\ell, \phi_\ell) \in (\Omega \otimes \Phi)$ such that one can say that $(\omega_k, \phi_k) \in (\Omega \otimes \Phi)$ is different from $(\omega_\ell, \phi_\ell) \in (\Omega \otimes \Phi)$ with a conclusion the (ω_k, ϕ_k) is a variety v_k and (ω_ℓ, ϕ_ℓ) is a variety v_ℓ.

Definition 2.3.2 (*Variety*)
A an element $v_k \in \mathbb{V}$ is said to be a variety if $v_k \in \mathbb{C}_k$ and there does not exist another category $\mathbb{C}_\ell$ such that $v_k \in \mathbb{C}_\ell$ with the condition that if $v_k \in \mathbb{C}_k \Rightarrow v_k \notin \mathbb{C}_\ell \forall \ell \neq k$ and that the corresponding characteristic-signal $\mathbb{Z}_k = (\mathbb{X}_k \otimes \mathbb{S}_k) = \{z_j = (x_j, s_j) | j \in \mathbb{J}_k \subset \mathbb{J}^\infty, k \in \mathbb{L}^\infty\}$ of $v_k \in \mathbb{V}$ is not equal to $\mathbb{Z}_\ell = (\mathbb{X}_\ell \otimes \mathbb{S}_\ell) = \{z_j = (x_j, s_j) | j \in \mathbb{J}_\ell \subset \mathbb{J}^\infty, \ell \in \mathbb{L}^\infty\}$ which is the characteristic-signal disposition that covers a variety $v_\ell \in \mathbb{V}$.

Note: On Category and Variety
Every category is simply, a categorial variety. Every categorial variety presents different information from without, that is an *inter-categorial information* that induces an inter-categorial varieties of qualitative differences and the same information from within, that is *intra-categorial information* that induces sameness and quantitative categorial varieties in the matter-energy space. The inter-categorial information and intra-categorial information may help to examine the nature of variety distribution in the universe around matter-energy structures seen in terms of universal variety-set in all actual-potential polarities under relational continuum and universal unity. The definitions of category, variety and categorial variety are such that if we have the variety space, $\mathbb{V}$ that is partitioned into categorial varieties such that for any two elements $v_k, v_\ell \in \mathbb{V}$ if $v_k \in \mathbb{C}_k$ and $v_\ell \in \mathbb{C}_k$ then one may conclude that $v_k \in \mathbb{V}$ and $v_\ell \in \mathbb{V}$ are the same variety otherwise they constitute different varieties. The formation of categories, given the solution to the identification problem of varieties is to determine the subscript (k) or (ℓ). This may be done if the universal object set Ω is defined as a fuzzy space with a membership characteristic function $\mu_\Omega(v)$ over the epistemological space. In this way every fix-level (α-level) will make $k = f(\alpha)$ that will determine all the partitioned subscripts of the categories.

It may be kept in mind that there are two things taking place with the phenomenon of information. One is the content of information and the other is the communication or transmission of the content. The content of information as described by a characteristic disposition establishes the variety. The communication of the content is undertaken though the signal disposition that reveals the nature of the variety for informing, knowing, learning and teaching. This content-communication process of information is general to matter and energy, and it is this generality that is established by characteristic-signal disposition just like the general defining concepts of matter and energy. It is important to note that matter and energy as two universal categories can only be distinguished by their corresponding characteristic-signal dispositions. In other words, matter can be distinguished from energy as universal varieties by their information content. There are general definitions of contents and phenomenon of matter and energy and there are specific definitions of specific forms of matter and energy that relate to the concept of variety.

2.4 Simple Analytics of Structures of Category, Variety, Universal Space, Characteristic-Signal Disposition and Their Relationships

A category $\mathbb{C}$ is thus a collection of objects with the same *characteristics set* $\mathbb{X}_{\mathbb{C}}$ that is specified by the same characteristic-signal disposition under the principle of quantitative-qualitative dispositions. Every category is a collection of the same variety where different varieties belong to different categories. Given the characteristics set $\mathbb{X}$ the set $\mathbb{X}_{\mathbb{C}} \subset \mathbb{X}$ is such that $\left(\bigcup_{i \in \mathbb{I}^{\infty}} \mathbb{X}_{\mathbb{C}_i} = \mathbb{X}\right)$. Furthermore, $\left(\bigcup_{i \in \mathbb{I}^{\infty}} \mathbb{C}_i = \mathfrak{U}\right)$, where $\mathfrak{U}$ is the universal space which is infinite and $i \in \mathbb{I}^{\infty}$ is a generic element with $\mathbb{I}^{\infty}$ as defining an infinite index set of categories. The universal space $\mathfrak{U}$ is the collections of all objects of varieties and hence $\left(\bigcup_{i \in \mathbb{I}^{\infty}} \mathbb{C}_i = \mathfrak{U} = \mathbb{V}\right)$. The universal object set Ω, the universe $\mathfrak{U}$ and the variety space $\mathbb{V}$ are partitions with respect to $\mathbb{C}$-categories such that $\mathbb{C}_\ell \neq \varnothing$, $\bigcap_{\ell \in \mathbb{L}^{\infty}} \mathbb{C}_\ell = \varnothing, \forall \ell$, and for any $i \neq j \in \mathbb{L}^{\infty}$, $\mathbb{C}_i \cap \mathbb{C}_j = \varnothing$ with $\left(\bigcup_{\ell \in \mathbb{L}^{\infty}} \mathbb{C}_\ell = \Omega = \mathfrak{U} = \mathbb{V}\right)$. The matter-energy universal system is such that every category is non-empty and the emptiness of intersections allows identification which may or may not be fuzzy. The emphasis of the intersection is the cognitive distinction while the fuzziness is the ontological connectivity.

It may be noticed that corresponding to each element, $\omega_\ell \in \Omega$ and a phenomenon $\phi_\ell \in \Phi$ (that is $(\omega_\ell, \phi_\ell) \in (\Omega \otimes \Phi)$) there is a set of attributes, $\mathbb{X}_\ell$, a set of signals $\mathbb{S}_\ell$ and information $\mathbb{Z}_\ell = (\mathbb{X}_\ell \otimes \mathbb{S}_\ell)$ that identify it to create a variety $v_\ell \in \mathbb{V}$ which also belongs to a category $\mathbb{C}_\ell$. The collection of all, $\omega_\ell \in \Omega$ with attributes, $\mathbb{X}_\ell$ and signals $\mathbb{S}_\ell$ constitutes a category $\mathbb{C}_\ell$ of the same variety and the collection of all these categories constitutes the object universe $\mathfrak{U}$. The universal object set, Ω, is simply the object universe, $\mathfrak{U}$, without the defining characteristic set, $\mathbb{X}_\ell$, signal set $\mathbb{S}_\ell$, and information $\mathbb{Z}_\ell = \mathbb{X}_\ell \otimes \mathbb{S}_\ell$ that partitions, Ω into categories of entities in differentiations to define the space of varieties, where $\mathbb{Z}_\ell$ constitutes the conditions of identification. Thus $\Omega = \{\mathbb{C}_\ell | \ell \in \mathbb{L}^\infty\} = \{(\omega_\ell, \mathbb{X}_\ell, \mathbb{S}_\ell) | \ell \in \mathbb{L}^\infty\}$ where $\mathbb{X} = \bigcup_{\ell \in \mathbb{J}^\infty} \mathbb{X}_\ell$, $\mathbb{S} = \bigcup_{\ell \in \mathbb{J}^\infty} \mathbb{S}_\ell$, $\mathbb{J}^\infty = \bigcup_{\ell \in \mathbb{L}^\infty} \mathbb{J}_\ell$ and $\#\mathbb{X}_\ell = \#\mathbb{J}_\ell = \#\mathbb{S}_\ell$. Both $\mathfrak{U}$ and Ω represent the collection of the *primary categories of reality*. They constitute the potential space which are the collection of all the potential varieties that are distinguished from the space of the actual known and unknown varieties The Definitions specify the objective existence of entities, states, processes and events that constitute the sources and destinations of *characteristic-signal-based information set* which we have referred to as the objective information in terms of characteristic-signal disposition of the form, $\mathbb{Z}_\ell = (\mathbb{X}_\ell \otimes \mathbb{S}_\ell)$.

It is useful to keep in mind that the varieties and categorial varieties apply to entities, states, processes and events. The concept of event is future-time defined relative to past-present structures and is not directly relate to information deficiency over the epistemological space as has been explained in [R17.17]. In other words, events in general are equipped with forward time but not necessarily equipped with measures of uncertainty. The time-dependent concept of event applies to both ontological and epistemological spaces and are independent of awareness of ontological agents. It is only when demand of awareness of cognitive agents is introduced that one may define an event in relation to uncertainty with some specified measure. The time-dependent concept of event will be useful in the development of the theory of info-dynamics. The uncertainty-dependent concept of event will be useful when cognitive awareness is implied over the conditions of informing, knowing, teaching and learning, all of which will include conditions of source-destination communication as seen in the degrees of perfection. The information sources present four important items of the universal object set Ω, a universal set of phenomena Φ, the universal characteristics set $\mathbb{X}$ and the universal signal set $\mathbb{S}$ on the basis of which information is constructed to place varieties. In the variety-transformation processes one moves from potential space to the space of the actual without a passage through the possibility and probability spaces. In the knowing processes, however, one moves from the potential space $\mathfrak{U}$ to the possibility space $\mathfrak{P}$ and fro the possibility space to the probability space $\mathfrak{B}$ and to the space of actual varieties $\mathfrak{A}$.

In this epistemic frame, it is useful to speak of primary category of existence and derived categories of existence from the primary category. The universal object set is the *primary category of existence* abstracted on the basis of the universal characteristic and attribute sets that reside in the objects and provide a process of knowing through derivatives by methodological constructionism and reductionism over the epistemological space. The four sets $(\mathfrak{U}, \mathfrak{A}, \mathbb{X}, \mathbb{S})$ are considered as the *factual reality* within the development of the general information definition (GID) in the sense that the existence of the four sets is independent of the awareness of any object in the universal object set. The process link between the potential and actual spaces is through the characteristic-signal dispositions without any reference to the cognitive behavior over the epistemological space. The cognitive awareness link between the potential and actual spaces is also through the characteristic-signal disposition passing through the possibility and probability spaces of knowing. Thus the characteristic-signal-based approach for information the definition is the most general. This general definition is the expression of the *objective reality* about the universal objects of differential varieties. Furthermore, the universal object set, constituting matter and energy in different varieties, is infinitely closed under *category formation* at any given cosmic or cosmological time. It is also closed under *continual transformation* where *categorial conversion* presents the necessary conditions and *Consciencism* through decision-choice processes presents the sufficient conditions for the directions and the nature of transformations in the creation and destruction of varieties under substitution-transformation principle connecting the past to the present and the present to the future. The specific inter-dependent roles played by characteristic and signal dispositions have been discussed in [R17.17].

The driving idea behind this epistemic framework is that the study of information is the study of matter-energy varieties, which offers the pathways to the general epistemology and specific areas of knowing in static and dynamic processes. Every characteristic-signal disposition presents itself in conditions of qualitative and quantitative dispositions. Similarly, every theory constructed is about categorial varieties, where the categorial varieties tend to influence the development and choice of methods and techniques in information representation and epistemic analyses. It is here, that explanatory science with the corresponding explanatory theories and prescriptive science with the corresponding prescriptive theories acquire important epistemic meanings and usefulness for thought and practice in the epistemological space where every activity is knowledge-decision-choice driven [R4.7, R4.10]. In a simple relational structure, explanlxatory science and theories are about the study of behaviors of varieties in time and over time, while prescriptive science and theories are about the study of how to bring about the destruction of old varieties and the creation of new varieties in the actual-potential spaces in time and over time under the universal substitution-transformation principle in matter and energy where the principle of opposites offers a path to transformation games in the socio-natural actual-potential polarities.

Chapter 3
Time Sets, Transformation Decisions and Socio-Natural Information Processes

Every transformation of a socio-natural object is a process that defines a path of states. The transformation process is energy-matter dependent to create either a quantitative or qualitative motion or a simultaneity of both which may be induced by either intentional or unintentional decision-choice actions at each transformation state. The transformation process either quantitatively or qualitatively conceived, relates to time where the time may be continuous or discrete. Each state of the object of transformation generates information that establishes its identity for knowing during acquaintance. In presenting information as generated by transformations of varieties, such information at each point of transformation must be related to time while the path of transformation must be related to a multiplicity of time. The time and multiplicity of time must have appropriate mathematical representations of time points, time and time set for the development of the theory of info-dynamics in either explanatory or prescriptive theoretical system for understanding and application. The following concept of group is a useful starting point to provide an algebraic structure for the development of the time set which will have the required properties for the analytical work of qualitative and quantitative motions. This algebraic structure is a semigroup which appears as a set with a binary operation. The point of critical understanding is that by examining the concept of transformation of varieties and categorial varieties, a time as the fourth dimension of the co-ordinates of the universal existence has been indirectly introduced. Thus, the universal existence is seen in terms of matter, energy, information and time which collectively determine varieties and categorial varieties in time and over time. The time dimension allows one to discuss connectivity and separation of the past, the present and the future. The epistemic geometry of the four dimensions of the universal existence is shown in Fig. 3.1.

The four dimensions of the ontological existence may be mapped into the epistemological space of universal knowing in terms of four relational structures of uncertainty and time in the quality-quantity dispositions. The epistemic geometry of the four relational structures of the universal knowing is shown in Fig. 3.2. It may be kept in mind that uncertainty is a time-dependent concept as well as

K.K. Dompere, *The Theory of Info-Dynamics: Rational Foundations of Information-Knowledge Dynamics*, Studies in Systems, Decision and Control 114, https://doi.org/10.1007/978-3-319-63853-9_3

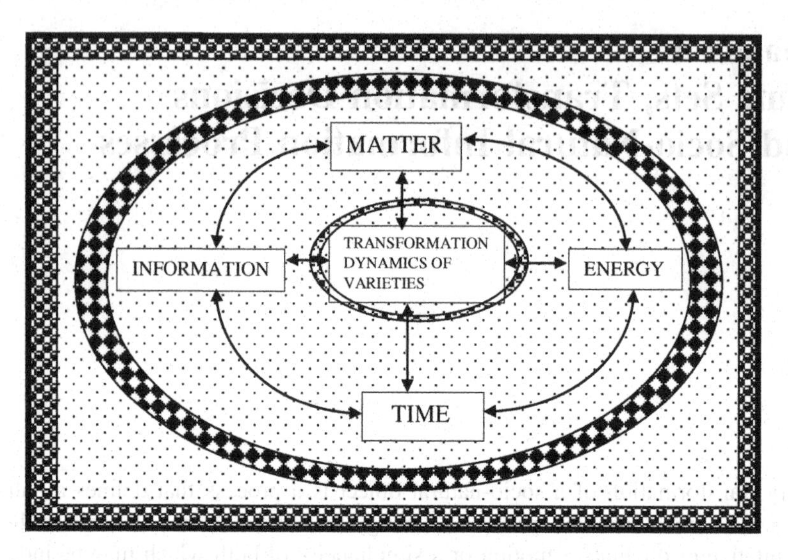

Fig. 3.1 An epistemic geometry of the four dimensions of the universal existence at the ontological space

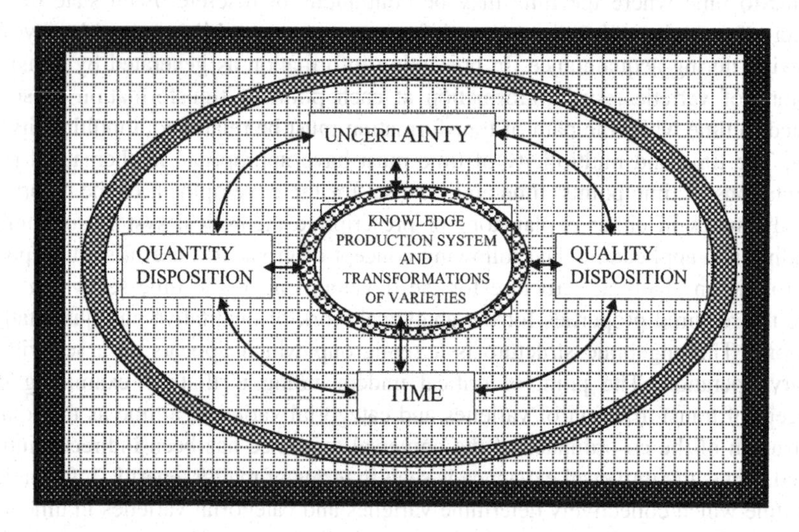

Fig. 3.2 An epistemic geometry of the four dimensions of the universal knowing at the epistemological space

decision-choice dependent concept in relation to events and outcomes. These events and outcome are themselves related to variety transformation that alter the characteristic-signal dispositions of elements separating them into differences and similarities through a defined knowing process.

3.1 Some Essential Definitions

The following definitions will be useful to the development of the needed mathematical structure of time-dependent process and processes.

Definition 3.1.1 *Semigroup*
A semigroup is a set $\mathbb{B}$ together with a function, +, that maps $\mathbb{B}^2$ into $\mathbb{B}$ (i.e., $+ : \mathbb{B} \otimes \mathbb{B} \to \mathbb{B}$) such that $b_0 + (b_1 + b_2) = (b_0 + b_1) + b_2, \forall b_0, b_1, b_2 \in \mathbb{B}$.

Definition 3.1.2 *Monoid*
A semigroup $\mathbb{B}$ is said to be a *monoid* if there is an element $0 \in \mathbb{B}$ such that $b + 0 = b = 0 + b$, $\forall b \in \mathbb{B}$. 0 is an *identity* and is unique under +.

Definition 3.1.3 *Division Relation*
A proper (improper) left division over a monoid, $\mathbb{B}$, is the relation, $<$ ($\leq$), defined on $\mathbb{B}$ such that $b_0 < b_1$ if $\exists\, b_2$ such that $b_2 \neq 0$ and $b_1 = b_0 + b_2$, $\forall b_0, b_1, b_2 \in \mathbb{B}$.
We shall now connect the monoid to the structure of time set, where the time set is shown to be a monoid and every element in the time set will be a time point. An important idea that may be kept in mind is that there is a pure time and a pure-time set, a transformation-decision time and transformation-decision time set, a cost time and a cost-time set, and a benefit time and a benefit-time set in the study of qualitative-quantitative dynamics of varieties and categorial varieties. Let us turn an attention to the construction of a pure time set relevant to the theoretical structure of info-dynamics as generated by transformation decisions without comparative cost-benefit considerations. The concepts of cost time, benefit time and cost-benefit time are more relevant with epistemological activities. They allow the time-preference order to be individually and collectively established by cognitive agents over the social transformation process, especially the preferred time for transformation actions including war and peace (For detailed discussions of these concepts and utilization see [R5.13], [R5.14], [R17.15], [R17.16].

3.2 A Construction and the Properties of a Pure Time Set, $\mathbb{T}$

Definition 3.2.1 If $\mathbb{T}$ is a monoid then $\mathbb{T}$ is a *time set* if and only if $\forall t_j \in \mathbb{T}$:

(R1) $\exists t_1$ and t_2 such that $t_1 = 0$ or $t_2 = 0$ and
(R2) $t_1 + t = t' + t_2$, $\forall t_1, t_2, t, t' \in \mathbb{T}$,
(R3) $t_1 + t = t + t_2 \Leftrightarrow t_1 = t_2$,
(R4) $t_1 + t = 0 \Rightarrow t_1 = 0$.

The definition of a time set as a monoid imposes an order relation on it. As a logical presentation, the time set may be viewed as a mathematical tool of the real time where $\mathbb{T} = (-\infty, +\infty)$. It may also be viewed as nested. Thus it must possess some useful and important properties for the study of transformations defined as inter-categorial and intra-categorial dynamics in the quantity-quality space, the phenomena of which may be partitioned into quantity-time phenomena and quality-time phenomena under the past-present-future structure such that for every present t_0 there is a past time t_1 and also there is a future time t_2 in such a way that $t_1 < t_0 \Rightarrow (t_1 - t_0) < 0$ and $t_2 > t_0 \Rightarrow (t_2 - t_0) > 0$ with $(t_2 - t_1) > 0 \Rightarrow t_1 < t_0 < t_2$. The value $\Delta t_1 = (t_1 - t_0)$ is the present-past time distance while the value $\Delta t_2 = (t_2 - t_0)$ is the present-future time distance while $\Delta t_3 = (t_2 - t_1)$ is past-future time (*sanakofa-anoma*) distance.

NOTE

The condition of nestedness implies that the time set may be related to the set of real numbers on the real line. Thus given $\tau_1 \leq \tau_2 \leq \cdots \tau_k \leq \cdots t_k \leq t_2 \leq t_1 \cdots$, the order behavior satisfies the following set containment relations; $[\tau_1, t_1] \supseteq [\tau_2, t_2] \supseteq [\tau_3, t_3] \supseteq \cdots \supseteq [\tau_k, t_k] \supseteq \cdots$ or $[\tau_k, t_k] \subseteq \cdots \subseteq [\tau_3, t_3] \subseteq [\tau_2, t_2] \subseteq [\tau_1, t_1]$ such that the term $\bigcup_{k \in R^\infty} [\tau_k, t_k]$ provides a time-ordered interval of the form $(t_1 - \tau_1) \geq (t_2 - \tau_2) \geq (t_3 - \tau_3) \geq (t_k - \tau_k) \geq \cdots$

Definition 3.2.2 *Conditions of a Time Set*

A set, $\mathbb{T}$ is said to be a *time set* if:

(1) It is equipped with a function, +, such that whenever there exist t_1 and t_2 with either $t_1 = 0$ or $t_2 = 0$ then $t_1 + t = t' + t_2$ for t_1, t_2, t and $t' \in \mathbb{T}$ and;

$$t_1 + t = t + t_2, \quad \text{iff } t_1 = t_2;$$
$$t_1 + t = 0 \Rightarrow t_1 = 0.$$

(2) There is a *complementation function*, $(-)$, defined over $\mathbb{T}$ where $t' - t$ is defined if either $t = t'$ or $t < t'$ and $(t' - t) \in \mathbb{T}$

(3) The function $(+)$ and the complementation function $(-)$ allow the past-present-future connectivity relation to be defined over the time set $\mathbb{T}$ such that the forward telescopic picture provides a historic future while the backward telescopic picture provides the historic past relative to the historic present. The function $(+)$ is viewed as forward extension into the future relative to the present while the function $(-)$ is viewed as backward extension into the past relative to the present.

The definition of the time set imposes an order relation on how time is cardinally viewed in terms of past-present-future relation and transformation processes. As a logical representation, we can view the time set as a mathematical concept and tool of the real line. It may also be viewed in period terms as a nested set with desired properties that induce either connectedness or continuity. The forward time relation allows the study of the present to the future, the backward time relation allows the study of the present to the past while the identity allows the study of present or initial conditions. The three together allows the past-present-future connectivity of continual processes. The time set, $\mathbb{T}$, may be considered to have a one-to-one correspondence with either the set of nonnegative reals, $\mathbf{R}^+$, or the set of non-negative integers, $\mathbf{N}^+$, with a zero identity under addition.

The correspondence of the time set with the nonnegative reals creates the possibility of developing time-continuous processes while the one-to-one correspondence with the set of non-negative integers offers the possibility of developing time-discrete processes. In both cases, present and future become either time-continuous or time-point connected that will allow models of dynamic and static process to be constructed over both the ontological and epistemological spaces. Furthermore, all transformational events such as costs and benefits from the past to the future are time-ordered, where the past and the future are always viewed relative to a fixed-time point referred to as present in the set, $\mathbb{T}$. The complementation function allows the past and future to be connected to the present in a discrete or continuous way such that the past historical information can be modeled as time-ordered processes. This past-present-future connectivity in relation to information and decision-choice activities is the *Asantrofi-anoma* rationality in relation to the *Sankofa-anoma principle* on the conditions of the continuous time connectivity.

Theorem 3.2.1 *If a monoid, $\mathbb{T}$, is a time set then the following properties hold for all $t \in \mathbb{T}$*

(i)	$t_1 + t = t + t_1;\ t, t_1 \in \mathbb{T}$	(*commutativity*),
(ii)	$t + t_1 = t + t_2 \Rightarrow t_1 = t_2$	(*left cancellation*),
(iii)	$t < t_1$ or $t = t_1$ or $t_1 < t$	(*connectedness*),
(iv)	$t \nless t$	(*irreflexivity*),
(v)	$t < t_1 \Rightarrow t_1 > t$ (i.e., $t < t_1 \Rightarrow t_1 \nless t$)	(*asymmetry*),
(vi)	$t < t_1$ and $t_1 < t_2 \Leftrightarrow t < t_2$	(*transitivity*),
(vii)	$t < (>) t_1 \Leftrightarrow t_2 + t < (>) t_2 + t_1$	(*left(right)invariance*),
(viii)	$t < t_1 \Rightarrow t < t_1 + t_2,\ t_2 \geq 0$	(*right extension*),
(ix)	$0 < t \Leftrightarrow t \neq 0$	(*least element*)
(x)	$t_1 \neq 0 \Leftrightarrow t < t + t_1$	(*continuity*).

Proof

(i) If $t_1 = t_1$ then $t_1 + t = t + t_1$ (by Definition 3.2.1).
If $t + t_1 = t + t_2$ then $t_1 + t = t + t_2$ (by (i), Theorem 3.2.1). This implies $t_1 = t_2$ (by Definition 3.2.1).
Since $\mathbb{T}$ is a time set $\exists t_1$ and t_2 such that

$(t_1 = 0 \text{ or } t_2 = 0)$ and $t_1 + t = t' + t_2$ (Definition 3.2.1)
then $(t_1 \neq 0 \text{ and } t_2 = 0 \text{ and } t_1 + t = t' + t_2)$ or $(t_1 = 0 \text{ and } t_2 = 0 \text{ and } t_1 + t = t' + t_2)$
$(t_1 = 0 \text{ and } t_2 = 0 \text{ and } t_1 + t = t' + t_2)$ so that
$(t_1 \neq 0 \text{ and } t' = t + t_1 \text{ or } t = t')$ or $(t_2 = 0 \text{ and } t = t' + t_2 [\text{by i}])$

Thus $t < t'$ or $t = t'$ or $t' < t$.
If $t_1 \neq 0$ and $t = t + t_1$ then $t_1 = 0$ and hence $0 + t = t + t_1$ So that $t_1 \neq 0$ by assumption and $t_1 = 0$ by deduction, resulting in a contradiction; thus $t < t$
Suppose the asymmetry is not true. Then for

$$t_1 \neq 0, t' = t + t_1 \text{ and for } t_2 \neq 0, t = t' + t_2$$

it follows that $t_1 \neq 0$ and $t_2 \neq 0$ and $t = (t + t_1) + t_2$. Then $t_1 + t_2 \neq 0$ and $t = t + (t_1 + t_2)$,
(by Definition 3.2.1) implies $t < t$, contradicting (iv).

(vi) For the transitivity property let $t_1 \neq 0$ and $t' = t + t_1$, and $t_2 \neq 0$ and $t'' = t' + t_2$.
Then $t_1 \neq 0$ and $t_2 \neq 0$, and $t'' = (t + t_1) + t_2$..
So that $(t_1 + t_2) \neq 0$ and $t'' = t + (t_1 + t_2)$
Since $t < t'$, $t' < t''$ and $t_1 + t_2 \neq 0$

(vii) Consider t, t_1 and $t_2 \in \mathbb{T}$ and $t < t_1$ but $t_2 + t \nless t_2 + t_1$.
Now $t' \neq 0$ and $t_1 = t + t'$ iff $t' \neq 0$
and $t_2 + t_1 = t_2 + (t + t')$ iff $t' \neq 0$ and $t_2 + t_1 = (t_2 + t) + t'$ (by ii) since $t' \neq 0$
This yields $t_2 + t < t_2 + t_1$ (a contradition),
hence $t_2 + t < t_2 + t_1$.

(viii) Consider $t, t_1, t_2 \in \mathbb{T}$ and $t < t_1$ and $t < t_1 + t_2$.. Let $t' \neq 0$ and $t_1 = t + t'$.
Thus $t' \neq 0$ and $t_1 + t_2 = (t + t') + t_2$.
Then $t' + t_2 \neq 0$ and $t_1 + t_2 = t + (t' + t_2)$; (by R3 of Definition 3.2.1).
Hence $t < t_1 + t_2$.

(ix) Let $t \in \mathbb{T}$ and $t \neq 0$. Then $t = 0 + t$ ($\mathbb{T}$ as a monoid), hence $0 < t$.
If $t = 0$ then $0 \nless 0$ (by iv), thus $0 < t$ iff $t \neq 0$.
Let t_1 and $t \in \mathbb{T}$. We have $t_1 \neq 0$ iff $0 < t_1$.
Thus $t + 0 < t + t_1$ and $t < t + t_1$ (by vii and ix). □

The time set, $\mathbb{T}$, with the properties of Theorem 3.2.1 may have a one-to-one correspondence with either the set of nonnegative reals, $\mathbb{R}^+$, or the set of non-negative integers, $\mathbf{N}^+$, with an identity 0 under addition. In both cases the left division is the strict inequality while (= , <) allow either connectedness or continuity to be defined on $\mathbb{T} = (-\infty, +\infty)$. Ones again the correspondence with the nonnegative reals creates the possibility of developing time-continuous processes while the one-to-one correspondence with the nonnegative integers offers the possibility of developing time-discrete processes in information production generated by variety transformations. In both cases, present and future become either time continuous or time-point connected. Furthermore, a complementation function $(-)$ is definable on $\mathbb{T}$ such that if $t_1 < t_2$ and $t_1, t_2 \in \mathbb{T}$ then $(t_2 - t_1) \in \mathbb{T}$. The complementation function allows the past to be studied as a process and to be connected to the present and linked to the future in the information generating processes of the transformation decisions. In the transformation decision every pure time point is defined by cost-benefit time which establishes the incentive to act.

We must add that the cost-benefit time reconciles the conflicts of incentives of transformations in the actual-potential polarity. The distribution of the cost-benefit time set over the relevant sub-set of the time set will define the path of the transformation that provides equivalent values for information aggregation. The continuity of time is seen as the *connectivity problem* in the time set, $\mathbb{T}$ which is handled by the left and right cancellations as a past-present-future relation in the *sankofa-anoma* tradition. We may then proceed with the following definitions. The time-connectivity problem is then linked to the solution to transformation-connectivity problem around the dynamics of quantity-quality varieties and categorial varieties for the understanding of inter-categorial and intra-categorial conversions of varieties. It is useful to keep in mind that inter-categorial conversion is qualitative dynamics and intra-categorial conversion is quantitative dynamics where the time dimension is neutral to the behaviors of matter, energy and information. The neutrality of time in transformations is also the passivity of time in all the processes.

Definition 3.2.3 *Time Continuity*

The continuity of time is a connectivity relation $(\sim)$ on the time set $\mathbb{T}$ such that any two time points $(\tau, \mathrm{t} \in \mathbb{T})$ are said to be connected and continuous if there is a connecting time subset $\mathbb{T}_{\mathbf{c}}$ with $(\mathbb{T}_{\mathbf{c}} \subset \mathbb{T}) \ni [\tau, \mathrm{t}] \subseteq \mathbb{T}_{\mathbf{c}}$ and the connectivity of the time points are written as $\tau_k \sim \mathrm{t_j}, \forall k, \mathrm{j} \in \mathbf{R}^+$

Proposition 3.2.1 Connectivity Relation

If $\mathbb{T}$ is a continuous real time set, then the connectivity relation is an equivalent relation in the sense that it is reflexive, symmetric and transitive. That is:

1. *Conditions of reflexivity*: $\forall \tau_k, \mathrm{t_j} \in \mathbb{T}$ and $\forall k, \mathrm{j} \in \mathbf{R}^+ \ \tau_k \sim \tau_k$ and $\mathrm{t_j} \sim \mathrm{t_j}$ since $\mathbb{T}_{\mathbf{c}}(\tau_k)$ is self-connected and $\mathbb{T}_{\mathbf{c}}(\mathrm{t_j})$ is also self-connected to produce reflexivity condition in the pure time set.
2. *Conditions of symmetry*: is such that $\forall \tau_k \mathrm{t_j} \in \mathbb{T}$ and $\forall k, \mathrm{j} \in \mathbf{R}^+, \tau_k \sim \mathrm{t_j} \Leftrightarrow \mathrm{t_j} \sim \tau_k$ which simply means that $[\tau_k, \mathrm{t_j}] \subseteq \mathbb{T}, \forall k, \mathrm{j} \in \mathbf{R}^+$ in that both time points are connected and hence continuous.
3. *Conditions of transitivity*: $\forall \tau_k \mathrm{t_j} \mathrm{t}_\ell \in \mathbb{T}$ and $\forall k, \mathrm{j}, \ell \in \mathbf{R}^+ (\tau_k \sim \mathrm{t_j})$ and $(\mathrm{t_j} \sim \mathrm{t}_\ell) \Rightarrow (\tau_k \sim \mathrm{t}_\ell)$ which simply means that the union of connected time subsets such as $[\tau_k, \mathrm{t_j}] \subseteq \mathbb{T}_{\mathbf{c}}$ and a connected time subset $[\mathrm{t_j}, \mathrm{t}_\ell] \subseteq \mathbb{T}_{\mathbf{c}'}$ is such that $[\tau_k, \mathrm{t}_\ell] \subseteq \mathbb{T}_{\mathbf{c}} \bigcup \mathbb{T}_{\mathbf{c}} \subset \mathbb{T}$.

A Reflection 3.2.1
A real or pure time set $\mathbb{T}$ may be viewed as a collection of time points or time singletons that are connected. In this view, the time set is a composition of connected time singletons to establish a continual time paths for the study of transformation of varieties. The time singletons may be seen as point-decompositions of $\mathbb{T} = (-\infty, +\infty)$ that involves its connected subsets of the form $\mathbb{T}_{C_\ell} \forall \ell \in R^+ \bigcup R^-$. The time set of interest is $\mathbb{T} = \bigcup_{\ell \in R^+ \cup R^-} \mathbb{T}_{C_\ell}$, where $\mathbb{T}_{C_\ell} \neq \emptyset$. In the study of transformation-substitution decision-choice processes, within destruction-construction socio-natural actual-potential polarities under the dynamics of dualities to generate info-dynamics that affect the behavioral structure of information stock-flow dynamics, time-point and time-period are essential instruments of analytics that allow the past-present-future transformation connectivity. This is important to understand the analytics of socio-natural transformations in destructive-creative space under cost-benefit interactions and conditions.

Definition 3.2.4 *Time Period and time moment*
The time period is a decomposition of connected time subset $\mathbb{T}_c \subset \mathbb{T}$ such that if $[\tau_k, t_j] \subseteq \mathbb{T}_c$ then $(t_j - \tau_k) > 0$ where $\tau_k, t_j \in \mathbb{T}$. The time period is said to be a time moment τ_k if given another time element $t_j \in \mathbb{T}$ then $(t_j - \tau_k) = 0 \Rightarrow t_j = \tau_k$.

NOTE
The time connectivity relates to continuous transformation analytics in the domain of $[\tau_k, t_j] \subseteq \mathbb{T}_{C_\ell} \subset \mathbb{T}, \forall \ell \in R^+ \bigcup R^-$ while the time non-connectivity relates discrete transformation analytics under the conditions of $(t_j - \tau_k) = 0 \Rightarrow t_j = \tau_k$ with $t_j \not\sim \tau_k$. The connectivity of time and the continuous transformation are the natural order, while the non-connectivity of time and discrete transformations are analytical processes to understand the continual dynamics through the methods of discrete dynamics as applied to the behavioral conditions of varieties and categorial varieties in time and over time.

3.3 Constructions of Transformation-Decision Time, Cost Time and Benefit Time

The dynamics of information production may be viewed in two interdependent sets of transformation-decision actions. One set of the transformation-decision actions is natural and the other set of the transformation-decision actions is social. The behavior of both sets of transformation decisions are explained under the theory of *Philosophical Consciencism* which is subdivided into *natural Philosophical Consciencism* and *social Philosophical Consciencism* which are internal to the socio-natural elements. The theory of Philosophical Consciencism studies the *sufficient conditions* for the flow of information through internal qualitative and quantitative transformations. The natural Philosophical Consciencism is a social-intentionality free within the relevant transformation-decision time period while the social Philosophical Consciencism is social intentionality dependent within transformation-decision time period in the sense of the system under command and control towards a set of goals and objectives. Every variety $\mathrm{v}_i \in \mathbb{V}, i \in \mathbb{I}^\infty$ is associated with information set $\mathbb{Z}_i$, $i \in \mathbb{I}^\infty$ to establish its categorial identity. It takes time for a variety to be transformed from say $\mathrm{v}_i \in \mathbb{V}, i \in \mathbb{I}^\infty$ to $\mathfrak{w}_i \in \mathbb{V}, i \in \mathbb{I}^\infty$. Such a time period will be called *transformation-decision time* $\mathbb{T}_{\mathfrak{D}}$ (it is the time period for which the categorial transversality conditions are satisfied).

Let us consider an individual object which may be a primary variety of the form $\mathrm{v}_i \in \mathbb{V}, i \in \mathbb{I}^\infty$ or a derived variety of the form $\mathfrak{w}_i \in \mathbb{V}, i \in \mathbb{I}^\infty$. Let us keep in mind that every derived categorial variety also serves as an intermediate primary categorial variety for a subsequently derived categorial variety. This point will become clear in the discussions of the subsequent chapters. The connectivity and the continuity of time allow the development of information flows while the discreteness of time allows the examination of information stocks at each time point from the past to the future. The time connectivity, the continuous time and discrete time allow the critical examinations of information stock-flow conditions. Every transformation-decision time $\mathbb{T}_{\mathfrak{D}}$ for any defined variety is in duality of cost time and benefit time of any time point $\mathrm{t}_j \in \mathbb{T}$. The cost-benefit time is in a relational continuum and unity without which any transformation-decision is undefinable. In terms of transformation-decision and the process of categorial conversion, three time sets may be distinguished and constructed from the real time set, $\mathbb{T}$, by combining it with either a set of real costs, a set of real benefits or a set of net cost-benefit values. The two evaluation-time sets are called *cost time*, $\mathbb{T}_{\mathfrak{C}}$ and *benefit time* $\mathbb{T}_{\mathfrak{B}}$. The two sets combine ordinal and cardinal scales to define a new time scale in a fuzzy domain. This fuzzy domain is especially relevant in the social transformations where intentionality drives the transformation direction through penumbral regions of social activities. The fuzzy domain, however, is unimportant in natural transformation decisions since vagueness in not part of information in the ontological space.

3.3.1 Cost, Benefit and Transformation as Information Process

The transformation process is destruction-construction activities where old varieties are destroyed and new varieties are created as substitutes. The destruction of the old varieties constitutes the real costs for the creation of new varieties which constitute the benefits while the new varieties constitute the costs of destroying the benefits of the old varieties in dualistically transformation processes. This relational structure may be viewed in terms opportunity costs where the path of transformation is nothing more than the cost-benefit path for current varieties. At each transformation-decision path the existing variety serves as the benefit from the previous variety as well as a cost for the future variety viewed in a *problem-solution duality* in both socio-natural processes. Let the assessment of cost be a set of characteristics $\mathfrak{C}$ and the assessment of benefits be $\mathfrak{B}$ and the transformation decision be a set of actions or controllers of the form $\mathfrak{D}$ where all transformation decisions depend on the cost-benefit rationality of the form $\mathfrak{D} = \mathfrak{f}(\mathfrak{C} \otimes \mathfrak{B})$. It may be pointed out that information enveloping is a problem-solution enveloping. It is also cost-benefit enveloping all of which relate to variety transformations. In this process, one must distinguish between cost-benefit rationality and the *Asantofi-anoma* principle in transformation dynamics. The cost-benefit rationality states that one must balance real costs and real benefits of transformations of varieties and select the best relative cost-benefit variety among the set of possible varieties. The Asantrofi-anoma principle simply states that there cannot be a selection of the benefit without the corresponding cost of any alternative variety among the possible varieties for transformation. We now may define cost time, benefit time and cost-benefit time in transformation processes that generate information dynamics.

Definition 3.3.1.1 A *cost time set,* $\mathbb{T}_{\mathfrak{C}}$ is a collection of pairs of the form

$$\mathbb{T}_{\mathfrak{C}} = \{(t, \mathfrak{c}) | t \in \mathbb{T} \text{ and } \mathfrak{c} \in \mathfrak{C}\} = \mathbb{T} \otimes \mathfrak{C}$$

Where $\mathfrak{C}$ is a set of real cost values defined in terms of the characteristics of the relevant categorial variety, where $\mathfrak{c}$ is fixed in $\mathfrak{C}$ and t is fixed in $\mathbb{T}$ and $\otimes$ is a Cartesian product.

Definition 3.3.1.2 A *benefit time set* $\mathbb{T}_{\mathfrak{B}}$, is a collection of pairs of the form

$$\mathbb{T}_{\mathfrak{B}} = \{(t, \mathfrak{b}) \mid t \in \mathbb{T} \text{ and } \mathfrak{b} \in \mathfrak{B}\} = \mathbb{T} \otimes \mathfrak{B},$$

where $\mathfrak{B}$ is a set of benefit characteristics of the relevant primary categorial variety and $\mathfrak{b}$ is fixed in $\mathfrak{B}$ at a given t in $\mathbb{T}$.

In the natural transformation-decisions, it is useful that the nature values the cost characteristic and the benefit characteristics in the same way where there is no subjective evaluation but merely ontological transformations of identities. This cannot be said of social transformation-decisions where vagueness and cognitive limitations interact to define subjectivity of evaluation and intentionality of goal-objective set. At the level of social transformations of varieties, the valuation of benefit time is *benefit-induced* in the sense that cognitive agents are present-oriented if the present generates benefit characteristics for current enjoyment in accord with social subjective preferences. As such, the society assesses the present time as having more value than future time relative to benefits. The same society has a different valuation of the same time point. This time valuation is *cost-induced* in the sense that the society is future-oriented relative to cost conditions of transformation. The society prefers the present if either current benefits are higher or future costs are higher. The society, therefore, assesses the future time to be more (less) valuable relative to cost (benefit) considerations in terms of transformation-decision where existing varieties are destroyed and in place new varieties are created under the economic principle of transformation-substitution process. The combination of these two concepts of time constitutes what may be referred to as a *transformation-decision time* that has its corresponding *time set.*

Definition 3.3.1.3 A *transformation-decision time set* $\mathbb{T}_{\mathfrak{D}}$, is a triplet of the form

$$\mathbb{T}_{\mathfrak{D}} = \{(t, \mathfrak{b}, \mathfrak{c},)|t \in \mathbb{T},\ \mathfrak{b} \in \mathfrak{B},\ \text{and } \mathfrak{c} \in \mathfrak{C}\} = \mathbb{T} \otimes \mathfrak{B} \otimes \mathfrak{C}$$

where $\mathbb{T}$ is a time set, $\mathfrak{B}$ is a set of benefit characteristics and $\mathfrak{C}$ is a set of cost characteristics associated with $t \in \mathbb{T}$ for any categorial variety.

Any socio-natural decision time point is composed of cost and benefit considerations of characteristics and switching of characteristics through a destruction-construction process of a continual destruction of existing varieties in the space of the actual and creation of varieties from the potential space under the transformation-substitution principle to generate a continual information stock-flow processes without end. On the benefit scale of social transformation-decision time, the present is more valuable than the future while on the cost scale of the same transformation-decision time the future is more valuable than the present as subjectively valued by cognitive agents in terms of relative cost-benefit conditions where cost produces undesirable (pain) effects and benefit produces desirable (pleasurable) effects. It is these cost-benefit-time implications and their deferentially distributional effect over cognitive agents, supported by the finality of life and the fear of information-decision failure that help to explain resistances to social transformations. Cognitive agents within any given society, therefore, face never-ending conflicts in terms of relative degrees of preference for any time point that is characterized by both costs and benefits associated with transformation-decisions over socio-natural varieties within the cost-benefit duality.

The analytical structure reveals distributions and continually changing distributions of cognitively conceptual dualities and polarities that drive the dynamic behavior of the universal problem-solution duality to a never-ending information stock-flow disequilibrium process. The distribution of cognitive dualities and assessments of categorial dynamics of actual-potential polarities are essential to the understanding of the conceptual domain of info-dynamics. The concepts of real time, cost, cost time, benefit and benefit time are of different structures over the ontological-decision space and over the epistemological-decision space. The cost-time set and the benefit-time set are distributions over the real time line and distinguished by preference ordering to establish preferential varieties with corresponding identities seen in terms of characteristic-signal dispositions. The cost-benefit preferences are established over the distribution of the time-point characteristic dispositions which are revealed by the distribution of the time-point signal dispositions over the epistemological space. One thing that is common to them is that the actual is always the cost of the potential in the methodological constructionism while the potential is always the benefit that fulfils the substitution principle and the principles of non-destructibility of matter and energy for matter-energy stock-flow equilibrium processes in the socio-natural transformation processes to generate new information for info-stock-flow disequilibrium. Similarly, the potential is always the cost of the actual in methodological reductionism while the actual is always the benefit that fulfils the substitution principle and the principles of non-destructibility of matter and energy in the socio-natural transformation processes to generate and maintain the historical conditions of the past. The conditions of matter-energy non-destructibility finds explanation in transformation-substitution principle of categorial dynamics of varieties.

3.3.2 Information Relativity of Transformation-Decision Time, Cost Time and Benefit Time

Every primary or a derived variety $v \in \mathbb{V}$ sits on cost-benefit duality defined by opposing negative-positive characteristic subsets that present the identity of the variety within the dynamics of actual-potential polarity for continual transformation. The set of the cost characteristics corresponds to the set of negative characteristics and is defined by the cost information set $\mathbb{Z}_{\mathfrak{C}}$. The set of the benefit characteristics corresponds to the set of positive characteristics and is defined by the benefit information set $\mathbb{Z}_{\mathfrak{B}}$. Let the cost-benefit configuration be defined by a relative value $\Xi = (\#\mathfrak{B}/\#\mathfrak{C})$ with relative information support of the form $\Xi = (\#\mathbb{Z}_{\mathfrak{B}}/\#\mathbb{Z}_{\mathfrak{C}})$. Any transformation of actual variety loses the actual cost-benefit configuration $\Xi_{\mathfrak{A}}$, the corresponding actual relative information support $\Xi_{\mathfrak{A}} = (\#\mathbb{Z}_{\mathfrak{B}}/\#\mathbb{Z}_{\mathfrak{C}})_{\mathfrak{A}}$, however it is retained and not destroyed in the information-stock accumulation process. The transformation system acquires as a replacement the potential cost-benefit configuration $\Xi_{\mathfrak{P}}$ from the actualized variety with a new relative information support of the form $\Xi_{\mathfrak{P}} = (\#\mathbb{Z}_{\mathfrak{B}}/\#\mathbb{Z}_{\mathfrak{C}})_{\mathfrak{P}}$. The new information of the new variety is a flow that goes to update the information stock in a never-ending process.

The comparative analysis of $\Xi_{\mathfrak{A}}$ and $\Xi_{\mathfrak{P}}$ does not arise in the ontological transformation activities since all transformation-decisions overs varieties in the ontological space are viewed and must be viewed as identities in the sense of *what there is*. In this respect, cost time, benefit time cost-benefit time and the corresponding sets do not arise since ontological transformations are devoid of intentionality of ontological decision-choice processes. Things are different, however, over the epistemological space where epistemological transformations of varieties (called brand-names) are social and the corresponding decisions are induced by cognitive agents with *intentionalities.* Two things are altered from the conditions of the ontological space to conditions of epistemological space. They are the cost-benefit values of transformation with relative cost-benefit values of transformation-decision time-point, and the comparative analysis of cost-benefit configuration with the corresponding supporting relative information. The transformation decisions will be based on the relative cost-benefit values of the actual $\Xi_{\mathfrak{A}}$

and the potential $\Xi_{\mathfrak{P}}$ with the corresponding relative value of information relation of the actual $\mathfrak{S}_{\mathfrak{A}}$ and the potential $\mathfrak{S}_{\mathfrak{P}}$. The comparative relational structures of information and cost-benefit conditions are further complicated by the social transformation-decision time point $\mathbb{T}_{\mathfrak{D}}$ assessed in terms of relative value of benefit-time set $\mathbb{T}_{\mathfrak{B}}$ and cost-time set $\mathbb{T}_{\mathfrak{C}}$ that must have a defined *connectivity principle* with a relational continuum and unity of social transformations. For epistemic clarity, it is important to maintain information conditions of distinction between the ontological and epistemological spaces relative to varieties and transformation processes.

3.3.2.1 Connectivity Principle, Information Relativity and Transformation Decision

The relative cost-benefit conditions of the actual $\Xi_{\mathfrak{A}}$ may be compared to the relative cost-benefit conditions of the potential $\Xi_{\mathfrak{P}}$ in the epistemological space by cognitive agents in order to undertake a destruction decision of the actual variety and create a vacuum for a replacement of a potential variety. Let us keep in mind that there is always one actual variety to be destroyed and technically there are infinite potential varieties one of which may be actualized. In the transformation decision within the actual-potential polarity, two immediate steps are required by the cognitive agents. The first step is to construct a finite feasible preference set of potential varieties $\mathfrak{V} \subset \mathbb{V}$ from the infinite set of potential varieties $\mathbb{V}$. The second step is to order the potential varieties in the feasible preference set relative to the actual variety.

The construct of $\mathfrak{V} \subset \mathbb{V}$ may be done with *categorial indicator function* where the feasible preference set of potential varieties and its construct in addition to membership ranking require that the information on the selected potential varieties $\mathfrak{V} \subset \mathbb{V}$ must be linked to the present by some connectivity relational ranking factor for ordering of the feasible varieties for the transformation decision. This actual-potential connectivity process is what is called discounting ranking relational factor to bring the future time dependent cost-benefit information of the preferred potential varieties as an input into the present transformation decision. In other words, to compare the net benefit of the actual variety to the net benefit of an actualized potential variety from the feasible potential set. It must always be kept in mind that there is always one actual variety and there is analytically infinite potential varieties one of which may be actualized. There is always a small and limited set of potential varieties that may fulfill the feasibility conditions from which a selection may be made on the basis of cost-benefit rationality under the *asantrofi-anoma principle* where benefits and costs are simultaneous in every potential choice and that one cannot select the benefit and leave the cost. The corresponding decision-choice problem to the *asantrofi-anoma principle* is the *asantrofi-anoma problem* that gives an epistemic justification to general cost-benefit analysis for all decisions.

It is useful to reiterate the *asantrofi-anoma* principle, problem and rationality. The asantrofi-anoma principle is one of the member of elements in the system of principles of opposites, where every socio-natural element has costs and benefit characteristics as its information identity. The problem is a decision-choice one in terms of how one decides on the choice when confronted with multiple elements. The rationality is simply, one cannot choose the benefit characteristics and leave the cost characteristics of any element, and hence one must choose benefit characteristics relative to cost characteristics in any decision-choice system. In terms of information support of dynamic decision-choice system, the *asantrofi-anoma* information requirement for decision-choice action are connected to the *sankofa-anoma* information conditions to generate the relevant information conditions for the present transformation decisions over the epistemological space. It is at this decision-time point that one deals with the present-future connectivity of information through discounting as well as present-past information connectivity through forecasting. The *discounting-ranking process* is a future-present phenomenon in a ranking relation relative to the set of cost-benefit conditions in social transformations. There is also the information connectivity process of the transformation decision from the past to the present called *forecasting process*, where the experience of the past cost-benefit information is brought to the present as an input to assist the discounting-ranking relation for the current transformation decision. For explicit definitions of discounting and forecasting and their relationship to information see the monograph on info-statics [R17.17].

The *information discounting-connectivity ranking factor* between the information on the actual variety and information on any potential variety $\mathrm{v} \in \mathfrak{P}$ may be computed as, $\mathfrak{r} = (\Xi_{\mathfrak{A}}/\Xi_{\mathfrak{P}}) = \left(\frac{\mathfrak{C}_{\mathfrak{P}}}{\mathfrak{C}_{\mathfrak{A}}}\right)\left(\frac{\mathfrak{B}_{\mathfrak{A}}}{\mathfrak{B}_{\mathfrak{P}}}\right)$ and supported by either an individual or collective subjective time preference order in order to offer a channel of information aggregation over the present-future profile of transformation [R5.13], [R5.14], [R17.17]. This discounting ranking process is supported by transformation time-point information relativity of the form $\eta = (\mathfrak{S}_{\mathfrak{A}}/\mathfrak{S}_{\mathfrak{P}}) = \left(\frac{\#\mathbb{Z}_{\mathfrak{C}_{\mathfrak{P}}}}{\#\mathbb{Z}_{\mathfrak{C}_{\mathfrak{A}}}}\right)\left(\frac{\#\mathbb{Z}_{\mathfrak{B}_{\mathfrak{A}}}}{\#\mathbb{Z}_{\mathfrak{B}_{\mathfrak{P}}}}\right)$. These information relativity concepts define connectivity-ranking relations that are partially order relation in the sense that they are reflexive, asymmetric and transitive. Both concepts and computable values are intended to rank the potential varieties relative to the existing actual variety for a transformation decision. Direct comparisons of the costs and benefits at different time-points in terms of the actual projected costs and benefits may yield wrong a social transformation strategy.

In the comparative analytics, the loss of the actual variety $\mathfrak{a} \in \mathfrak{A}$ by transformation leads to the loss of the actual benefit as well as the actual cost characteristics which are replaced by benefit and cost characteristics of an actualized potential $\mathrm{v} \in \mathfrak{P}$. This is the conditions of the *Asantrofi-anoma principle*, where costs and benefits are simultaneously contained in any decision choice element and where one cannot get rid of the cost of the actual and simply retain the benefit by the choice process of transformation decisions [R17.15], [R17.16], [R5.13]. The *Asantrofi-anoma principle* always works with cost-benefit rationality under all decision-choice systems. It is this principle that demands cost-benefit analysis with a defined cost-benefit rationality which provides a general framework of decision-choice science. It must be noted that the concept of opportunity cost is relative to benefit-cost conditions of varieties and that cost-benefit analysis is a general analytical as well as an epistemic framework of both theoretical and applied decision-choice analytics. It may be noted that all constrained optimization problems are easily transformed into benefit-cost or cost-benefit constrained process under a goal-objective information process. A cost resides in a benefit and a benefit resides in a cost within the cost-benefit duality of transformational decision-choice processes in the general information production. Everything exists as a variety. Every variety finds its identity in a negative-positive duality which finds expression in quality-quantity duality that exists in a cost-benefit duality in the dynamics of a socio-natural actual-potential polarity.

3.3.2.2 Information Relativity Relation, Categorial Indicator Function and Transformation—Decision Analytics

The needed information structures on the set of feasible preferred potential varieties relative to the actual variety have been discussed. The information relativity relation on cost-benefit structures of the feasible varieties provides a preference order relation of the form $(\succsim)$ as an increasing preference order relation or in a form $(\precsim)$ as a decreasing preference order relation in the increasing-decreasing duality which must have a desirable mathematical properties in their representations. The categorial indicator function which is of the form $\mathcal{I}_{\mathfrak{P}}$ must also have appropriate mathematical structure that will allow the set of the feasible potential varieties to be formed relative to an actual variety $(\mathrm{v}_{\mathfrak{a}} \in \mathfrak{A})$. The partial order relation $(\succsim)$ or $(\precsim)$ defined on relative cost-benefit conditions is such that the following proposition can be established. The differential information discounting-ranking factor may be used to rank the potential varieties that may be actualized. All cost-benefit phenomena, all decision-choice conflicts, all conditions of games of socio-natural polarities with residing negative-positive dualities are only understood for actions in terms of varieties and their identities defined by characteristic dispositions and revealed by the corresponding signal dispositions that present the general information structure in the quality-quantity space.

Proposition 3.2.1 Information Relativity Relation on Cost-benefit Decision.

If $\mathbb{R} = \left\{\mathrm{r}|\mathrm{r} = (\Xi_{\mathfrak{A}}/\Xi_{\mathfrak{P}}) = \left(\frac{\mathfrak{C}_{\mathfrak{P}}}{\mathfrak{C}_{\mathfrak{A}}}\right)\left(\frac{\mathfrak{B}_{\mathfrak{A}}}{\mathfrak{B}_{\mathfrak{P}}}\right) \forall \mathrm{v} \in \mathbb{V}_{\mathfrak{A}} \& \mathrm{v} \in \mathbb{V}_{\mathfrak{P}}\right\}$ *defines a set of cost-benefit connecting information discounting ranking factors from the set of potential varieties* $\mathfrak{P} \subset \mathbb{V}_{\mathfrak{P}}$ *relative to a particular actual* $\mathrm{v}_{\mathfrak{a}} \in \mathfrak{A}$ *in a real time-set transformation process, and* $(\succsim)$ *is a preference order relation defined over the feasible set of potential varieties* $\mathfrak{P}$ *then the cost-benefit information discounting connectivity relation is such that the following conditions hold.*

1. *Conditions of reflexivity*: $\forall \mathfrak{r}_i \in \mathbb{R} i \in \mathbf{R}^+$, $\mathfrak{r}_i \succsim \mathfrak{r}_i$ since $\mathbb{R}$ is self-connected in that $\forall \mathfrak{r}_i \in \mathbb{R}\, i \in \mathbf{R}^+ \mathfrak{r}_i \precsim \mathfrak{r}_i$ and every element in $\mathbb{R}$ is a self-produced reflexivity condition in the pure actual-potential discounting of relative cost-benefit information in transformations.
2. *Conditions of asymmetry*: $\forall \mathfrak{r}_i \mathfrak{r}_j \in \mathbb{R}$ and $\forall i, j \in \mathbf{R}^+ \mathfrak{r}_i \succsim \mathfrak{r}_j \Leftrightarrow \mathfrak{r}_j \not\succsim \mathfrak{r}_i$. This simply means that if given two potential varieties $\mathfrak{w}_i, \mathfrak{w}_j \in \mathfrak{P}$ with $i, j \in \mathbf{R}^+$ relative to the same actual variety $v_a \in \mathfrak{A}$ if the cost-benefit information discounting factor for $\mathfrak{w}_i \in \mathfrak{P}$ with $i, \in \mathbf{R}^+$ dominates $\mathfrak{w}_j \in \mathfrak{P}$ with $j \in \mathbf{R}^+$ relative to $v_a \in \mathfrak{A}$ then there cannot be a reversibility of the order of preference as seen in the process of preference potential variety transformation replacement process in the actual-potential polarity. If $\forall i, j \in \mathbf{R}^+ \mathfrak{r}_i \succsim \mathfrak{r}_j$ and $\mathfrak{r}_j \succsim \mathfrak{r}_i \Rightarrow \mathfrak{r}_i = \mathfrak{r}_j$ and $\mathfrak{r}_j \sim \mathfrak{r}_i$ in the potential variety ranking for social transformation decision.
3. *Conditions of transitivity*: $\forall \mathfrak{r}_i, \mathfrak{r}_j, \mathfrak{r}_k \in \mathbb{R}$ and $\forall i, j, k \in \mathbf{R}^+$ if $(\mathfrak{r}_i \succsim \mathfrak{r}_j)$ and $(\mathfrak{r}_j \succsim \mathfrak{r}_k) \Rightarrow (\mathfrak{r}_i \succsim \mathfrak{r}_k)$ which simply means that there is a relative cost-benefit information connectivity among different potential varieties where given if $\mathfrak{w}_i, \mathfrak{w}_j, \mathfrak{w}_k \in \mathfrak{P} \subset \mathbb{V}$ with $i, j, k \in \mathbf{R}^+$ then it is the case that $\mathfrak{w}_i \succsim \mathfrak{w}_j \succsim \mathfrak{w}_k, \forall i, j, k \in \mathbf{R}^+$ in social assessment in the transformation decision under the connected cost-benefit time set.

The cost-benefit variety ranking for actualization through social transformation decision has been defined over a feasible set of potential varieties $\mathfrak{P} \subset \mathbb{V}$, the manner in which this is constructed must be shown. Let us keep in mind that every actual variety $v_a \in \mathfrak{P} \subset \mathbb{V}$ also has a structure where $v_a \in \mathbb{C} \subset \mathbb{V}$. The technique and process of constructing $\mathbb{C}$ as categorial varieties also apply to the construction of the feasible set of potential varieties. This technique is through the *categorial indicator function* which will become analytically useful when the fuzzy indicator function is introduced and discussed in relation to possibility set and probability distribution.

Definition 3.2.1 *Categorial Indicator Function*
Given a category $\mathbb{C} \subset \Omega$ of varieties, the feasible set of potential varieties $\mathfrak{P} \subset \mathbb{C}$ around an actual variety $v_a \in \mathfrak{A}$ may be constructed with a *categorial indicator function* of the form $\mathcal{I}_{\mathfrak{P}}$ from the general categorial indicator $\mathcal{I}_{\mathbb{C}}(v), v \in \mathbb{C} \subset \Omega$. Let the information support of the potential varieties in $\mathbb{C}_\ell$ be $\mathbb{Z}_\ell = \mathbb{X}_\ell \otimes \mathbb{S}_\ell$. Let the degree of feasibility be defined by a subjective ranking of fuzzy numbers in terms of membership function of the form $\mu_{\mathbb{C}}(v) \in [0, 1]$ with information support $\mu_{\mathbb{Z}}(\mathbb{Z}_v) \in [0, 1]$ then the categorial indicator function may be constructed as:

$$\mathcal{I}_{\mathfrak{P}}(v) = \begin{cases} 1 \text{ if } v \in \mathfrak{P} \subset \mathbb{C} \subset \Omega, \text{ and } \mu_{\mathbb{C}}(v) > \alpha \in (\alpha, 1] \\ 0 \text{ if } v \notin \mathfrak{P} \text{ but } v \in \mathbb{C} \subset \Omega \text{ and } \mu_{\mathbb{C}}(v) \leq \alpha \in [0, \alpha] \end{cases}$$

With the following properties:

1. $\mathcal{I}_{\mathfrak{P}}(\mathrm{v}) = 1 - \mathcal{I}_{(\mathbb{C}/\mathfrak{P})}(\mathrm{v})$
2. If $\mathfrak{P}_i\, \mathfrak{P}_k \subset \mathbb{C}$ are two feasible sets of potential replacement varieties then:

$$\mathcal{I}_{\mathfrak{P}_i \bigcap \mathfrak{P}_k}(\mathrm{v}) = \min[\mathcal{I}_{\mathfrak{P}_i}(\mathrm{v}), \mathcal{I}_{\mathfrak{P}_k}(\mathrm{v})] = [\mathcal{I}_{\mathfrak{P}_i}(\mathrm{v})] \cdot [\mathcal{I}_{\mathfrak{P}_k}(\mathrm{v})]$$
$$\mathcal{I}_{\mathfrak{P}_i \bigcup \mathfrak{P}_k}(\mathrm{v}) = \max[\mathcal{I}_{\mathfrak{P}_i}(\mathrm{v}), \mathcal{I}_{\mathfrak{P}_k}(\mathrm{v})] = [\mathcal{I}_{\mathfrak{P}_i}(\mathrm{v})] = [\mathcal{I}_{\mathfrak{P}_k}(\mathrm{v})] - \min[\mathcal{I}_{\mathfrak{P}_i}(\mathrm{v}), I_{\mathfrak{P}_k}(\mathrm{v})]$$

3. $\mathcal{I}^n_{\mathfrak{P}}(\mathrm{v}) = I_{\mathfrak{P}}(\mathrm{v}), \forall n \in \mathbb{T}^+$

The categorial indicator function is distinguished from membership characteristic function which relates to conditions of degrees of exactness, subjectivity, quality, vagueness, approximation, linguistic variables and others useful in fuzzy logical analytics and paradigm of thought. The categorial indicator function defined in a fuzzy space becomes equipped with decision-choice function to deal with subjectivity in judgement. For more discussion on categorial indicator function see [R17.15]. The use of the categorial indicator function may be constrained by a membership characteristic function. In this respect, the feasible set of potential varieties $\mathfrak{P} \subset \mathbb{C}$ relative to replacement of any actual variety $\mathrm{v}_{\mathfrak{a}} \in \mathfrak{A}$ may be redefined in a fuzzy space by a membership characteristic function in the form $\tilde{\mathfrak{P}} = \{(\mathfrak{P}, \mu_{\mathfrak{P}}(\mathrm{v})) | \, \mu_{\mathfrak{P}}(v) \in [0, 1]\}$. Thus the categorial indicator function for substitution-transformation process of any actual-potential polarity in the information dynamics becomes equipped with a membership characteristic function in the degrees of either exactness or subjectivity under the *principle of acquaintance*. The exactness and subjectivity must be related to the capacity of a potential variety $(\mathrm{v}, \mu_{\mathfrak{P}}(\mathrm{v})) \in \tilde{\mathfrak{P}}$ fulfilling the transformation replacement of a social objective.

NOTE
The composition of the categories as has been discussed as part of the theory of info-statics may be constructed with categorial indicator function of the form, $\mathcal{I}_{\mathbb{C}_\ell}(\omega), \omega \in \mathbb{C}_\ell \subset \Omega$ to create fuzzy equivalence classes with membership function $\mu(\cdot)$ such that the members are indistinguishably described by the same intra-qualitative information $\mathbb{Z}_\ell = \mathbb{X}_\ell \otimes \mathbb{S}_\ell$. The notion of fuzzy equivalence classes are fuzzy categories in the sense that the crisp categories are formed by the use of methodology of *fuzzy decomposition* and *fixed-level cut*. The fixed-level values are subjectively determined acquaintance and may vary over different categorial varieties. The formation of epistemological categories may not correspond to the ontological categories in that if $\mathfrak{E} = \{\mathbb{C}_\ell \,|\, \ell \in \mathbb{I}_{\mathfrak{E}}\}$ is the collection of epistemological categories with an expanding finite index set of $\mathbb{I}_{\mathfrak{e}}$, and $\mathfrak{O} = \{\mathbb{C}_\ell \,|\, \ell \in \mathbb{I}^\infty_{\mathfrak{O}}\}$ is the collection of ontological categories with infinite index set $\mathbb{I}^\infty_{\mathfrak{E}}$ then $\mathbb{I}_{\mathfrak{E}} \subset \mathbb{I}^\infty_{\mathfrak{O}}$ and $\mathfrak{E} \subset \mathfrak{O}$. It is also to be observed that the epistemological categories are related to knowledge categories such that if $\mathfrak{K} = \{\mathbb{C}_\ell \,|\, \ell \in \mathbb{I}_{\mathfrak{K}}\}$ with $\mathbb{I}_{\mathfrak{K}}$ as its expanding finite index set then $\mathbb{I}_{\mathfrak{K}} \subset \mathbb{I}_{\mathfrak{E}}$ with $\mathfrak{K} \subset \mathfrak{E} \subset \mathfrak{O}$.

3.3.3 *Fuzzy Time Set*, Fuzzy *Cost*, *Fuzzy Benefit and Transformation-Decision Analytics*

Let us now turn our attention from pure time set to transformation-decision time set. The concept of transformation-decision time set may be seen in terms of ontological and epistemological transformations which generate continual information flow to update the time-point information stocks. In this epistemic framework, information accumulation and information flow are time dependent, where the particular concepts of time and information define an important framework for attempts of cognitive agents to inform, learn and understand the universal system and its continual transformation. The *ontological transformation-decision time set* presents the concept of time identity in terms of the primary category of time set in ontological transformations of varieties and the generation of information flows in the ontological space. The *epistemological transformation-decision time set* presents the concept of a derived category of time set relative to the primary category of time set in epistemological transformations of varieties and the generation of information flows in the epistemological space. One may think of the ontological transformation-decision time set in relation to an *objective time set* relative to the pure time set and the epistemological transformation-decision time set as a *subjective time set* relative to the pure time set.

The subjectivity of the epistemological transformation-decision time set is in relation to vagueness that generates qualitative-quantitative characteristics in the time points of epistemological transformation decisions. The vagueness in the concept that flows from qualitative characteristics and quantitative approximation from acquaintances allows one to speak of a *fuzzy time set* from which a *crisp time set* may be constructed. Every transformation-decision time point sits on a cost-benefit duality of transformation decision with a relative time preference for differential distributions of costs and benefits over time, and with differential preference weights for cost characteristics and benefit characteristics at the same time point as assessed by cognitive agents. The differential weights attached to cost unit and benefit unit at the same time point allow one to speak of cost time and benefit time relative to transformation decisions on epistemological varieties. It is useful, therefore, to define and explicate the concepts of cost time, benefit time, fuzzy time and transformation-decision time sets. The fuzzy time set and the transformation-decision time set must meet the conditions of cost-time and benefit-time sets.

It may be observed that the subjective evaluations of each transformation-decision time point based on inter-temporal preferences are fuzzy irrespective of whether benefits or costs are associated. In other words, the valuations of time are based on fuzzy preferences due to the vagueness in conceptualizing time with a linguistic variable such as past, present and future. In this section, since the ontological transformation-decision time set is taken as the identity and the primary category, we shall work with a conceptual framework where every construct of time set in the epistemological space may be viewed as a derivative from the primary. We shall be interested in advancing an epistemic framework for relating inter-temporal fuzzy preferences to epistemological transformation-decisions of epistemological categorial varieties. The introduction of fuzzy paradigm of transformation-decision analytics allows one to deal with subjective phenomenon of uncertainties due to vagueness and information limitations that are experienced through acquaintances with the characteristic-signal dispositions in the cost-benefit dualities within the dynamics of actual-potential polarity.

We consider an epistemological transformation decision of any potential social variety $\mathrm{v} \in \mathfrak{V}$ to actual social variety $\mathfrak{w} \in \mathfrak{V}$, where the society faces two fuzzy time sets. One fuzzy time set is a set of benefit-time values with a membership function that defines the inter-temporal intensities associated with a unit benefit characteristic at a given time and a set of benefit information characteristics. The other fuzzy time set is a set of cost-time values equipped with a corresponding membership function that defines inter-temporal intensities of unit cost characteristics over the entire spectrum of epistemological transformation-decision time set at a given set of information cost characteristics. The conceptual implication, here, is that for each epistemic category the information on any $\mathbb{C}_\ell \subset \mathfrak{C}$ may be represented as benefit information $\mathbb{Z}_\ell^{\mathfrak{B}} = \mathbb{X}_\ell^{\mathfrak{B}} \otimes \mathbb{S}_\ell^{\mathfrak{B}}$ and cost information $\mathbb{Z}_\ell^{\mathfrak{C}} = \mathbb{X}_\ell^{\mathfrak{C}} \otimes \mathbb{S}_\ell^{\mathfrak{C}}$ such that $\mathbb{Z}_\ell = \mathbb{Z}_\ell^{\mathfrak{C}} \bigcup \mathbb{Z}_\ell^{\mathfrak{B}} = (\mathbb{X}_\ell^{\mathfrak{C}} \otimes \mathbb{S}_\ell^{\mathfrak{C}}) \bigcup \left(\mathbb{X}_\ell^{\mathfrak{B}} \otimes \mathbb{S}_\ell^{\mathfrak{B}}\right)$. It is important that under the principle of opportunity cost within the cost-benefit duality, a cost is seen as a benefit and a benefit is seen as a cost in an inter-supportive relation as one examines the dynamics of actual-potential polarities. The implication here over the epistemological space is that $\mathbb{Z}_\ell^{\mathfrak{C}} \bigcap \mathbb{Z}_\ell^{\mathfrak{B}} = \varnothing$ and $\mathbb{Z}_\ell^{\mathfrak{C}} \bigcap \mathbb{Z}_\ell^{\mathfrak{B}} \neq \varnothing$ are decision-evaluative possibilities. Whether an information is a benefit or a cost depends on the nature and type of transformation decision.

3.3.3.1 Some Essential Definitions of Concepts of Time Over the Epistemological Space

For the development of related concepts in the understanding the behavior of stock-flow information process, the following working definitions of time sets are offered in fuzzy and non-fuzzy spaces of epistemic analytics. Here, the analytics is on the fuzzy time sets.

Definition 3.3.3.1.1 *A benefit fuzzy time set,* $\tilde{\mathbb{T}}_\mathfrak{b}$, is a triplet of a time set, $\mathbb{T}$, a benefit set, $\mathfrak{B}$, and a membership function, $\mu_{\tilde{\mathbb{T}}_\mathfrak{b}}(\cdot)$ that may be represented as:

$$\tilde{\mathbb{T}}_\mathfrak{b} = \left\{ t_\mathfrak{b} = \left(t, \mathfrak{b}, \mu_{\tilde{\mathbb{T}}_\mathfrak{b}}(t)\right) \mid t \in \mathbb{T}, \mathfrak{b} \in \mathfrak{B} \text{ and } \mu_{\tilde{\mathbb{T}}_\mathfrak{b}}(t) \in [0, 1] \right\}$$

with a membership function that is decreasing in $\mathbb{T}$.

In other words, a benefit fuzzy time set is a time set with a membership characteristic function that specifies the grades of preference assigned to the same unit of benefit over each point of time through the relevant spectrum of a transformation-decision time. The benefit fuzzy time set is said to be a crisp benefit time set if $\mu_{\tilde{\mathbb{T}}_{\mathfrak{b}}}(\cdot) = 1$ and hence $\mathbb{T}_{\mathfrak{b}} = \{t_{\mathfrak{b}} = (t, \mathfrak{b}, 1) | t \in \mathbb{T}, \mathfrak{b} \in \mathfrak{B} \text{ and } \mu_{\mathbb{T}_{\mathfrak{b}}}(t) = 1\}$. The decreasing membership function means that the same units of present benefits are assess higher than the future benefits in the sense that decision agents prefer current benefit characteristics to the same benefit characteristics at the future date. That is, if $(t_0 \leq t_1 \leq t_2 \leq \cdots \leq t_n) \in \mathbb{T}$ then $\mu_{\tilde{\mathbb{T}}_{\mathfrak{b}}}(t_0) \geq \mu_{\tilde{\mathbb{T}}_{\mathfrak{b}}}(t_1) \geq \mu_{\tilde{\mathbb{T}}_{\mathfrak{b}}}(t_2) \geq \cdots \geq \mu_{\tilde{\mathbb{T}}_{\mathfrak{b}}}(t_n)$. This condition provides a justification for reasons why interest rates are charged on current benefit of loan that will be paid at a future time. The real interest rate value is a compensation to equalize the value of future benefit with a current benefit [R5.14].

Definition 3.3.3.1.2 *A cost fuzzy time set,* $\tilde{\mathbb{T}}_{\mathfrak{c}}$ is a triplet of a time set $\mathbb{T}$, a cost set, $\mathfrak{C}$, and a membership characteristic function, $\mu_{\tilde{\mathbb{T}}_{\mathfrak{c}}}(\cdot)$, that may be represented as:

$$\tilde{\mathbb{T}}_{\mathfrak{c}} = \left\{t_{\mathfrak{c}} = \left(t, \mathfrak{c}, \mu_{\tilde{\mathbb{T}}_{\mathfrak{c}}}(\cdot)\right) | \, t \in \mathbb{T}, \mathfrak{c} \in \mathfrak{C} \text{ and }, \mu_{\tilde{\mathbb{T}}_{\mathfrak{c}}}(t) \in [0, 1]\right\}$$

with a membership function that is increasing in $\mathbb{T}$.

The fuzzy cost time set is similarly defined as the fuzzy benefit time set with an opposite curvature. In other words, a cost fuzzy time set is a time set with a membership characteristic function that specifies the grade of preference assigned to the same unit of cost characteristic over each point of time through the relevant spectrum of a transformation-decision time. The cost fuzzy time set is said to be a crisp benefit time set if $\mu_{\tilde{\mathbb{T}}_{\mathfrak{c}}}(\cdot) = 1$ and hence $\mathbb{T}_{\mathfrak{c}} = \{t_{\mathfrak{c}} = (t, \mathfrak{c}, 1) | t \in \mathbb{T}, \mathfrak{c} \in \mathfrak{C} \text{ and } \mu_{\mathbb{T}_{\mathfrak{c}}}(t) = 1\}$. The increasing membership function means that the same units of present real costs are assess lower than the future real cost units in the sense that decision agents prefer real future cost characteristics to the same real cost characteristics at the current date. That is, if $(t_0 \leq t_1 \leq t_2 \leq \cdots \leq t_n) \in \mathbb{T}$ then $\mu_{\tilde{\mathbb{T}}_{\mathfrak{c}}}(t_0) \leq \mu_{\tilde{\mathbb{T}}_{\mathfrak{c}}}(t_1) \leq \mu_{\tilde{\mathbb{T}}_{\mathfrak{c}}}(t_2) \leq \cdots \leq \mu_{\tilde{\mathbb{T}}_{\mathfrak{c}}}(t_n)$. The subjective assessment of cost-benefit time sets forms the foundation for discounting the future information to the present as an input into transformation-decisions of epistemological varieties. In the evaluative process, $\mu_{\mathbb{T}_{\mathfrak{b}}}(t) \neq \mu_{\mathbb{T}_{\mathfrak{c}}}(t), \forall t \in \mathbb{T}$. The differential time weights of the same unit of cost and benefit characteristics lead to the question whether two different discounting rates must be used in assessing benefit information and cost information as inputs into the transformation-decisions of epistemological varieties.

Definition 3.3.3.1.3 *A fuzzy transformation-decision time set,* $\tilde{\mathbb{T}}_{\mathfrak{d}}$, is composed of benefit and cost time sets which is defined as $\tilde{\mathbb{T}}_{\mathfrak{d}} = \tilde{\mathbb{T}}_{\mathfrak{b}} \otimes \tilde{\mathbb{T}}_{\mathfrak{c}}$. The non-fuzzy transformation-decision time is $\mathbb{T}_{\mathfrak{d}} = \mathbb{T}_{\mathfrak{b}} \otimes \mathbb{T}_{\mathfrak{c}}$.

In the analysis that follows, a benefit (cost) fuzzy time set is also viewed as a fuzzy benefit (cost) time set. A fuzzy transformation-decision time set is composed of cost and benefit values with corresponding grades of time preferences by means of which a transformation-decision may be attached to one unit of cost and benefit values measured on the same scale. The curvature of the membership functions of cost and benefit time sets are provided as axioms.

Axiom 3.3.3.1.1 *Ranking of Cost and Benefit Time Values*

Let $(\succ)$ and $(\prec)$ be an increasing and decreasing preference relations established over cost-benefit time sets $\tilde{\mathbb{T}}_{\mathfrak{b}}$ and $\tilde{\mathbb{T}}_{\mathfrak{c}}$ and $(>)(<)$ are greater or less than relations defined over the time set $(\mathbb{T} = (-\infty, +\infty))$. For any two time points t_i and t_j and for any society:

$$t_i \left\{ \begin{matrix} < \\ > \end{matrix} \right\} t_j \Rightarrow \left\{ \begin{matrix} \mu_{\mathbb{T}_{\mathfrak{b}}}(t_i) \left\{ \begin{matrix} \succ \\ \prec \end{matrix} \right\} \mu_{\mathbb{T}_{\mathfrak{b}}}(t_j) \\ \mu_{\mathbb{T}_{\mathfrak{c}}}(t_i) \left\{ \begin{matrix} \prec \\ \succ \end{matrix} \right\} \mu_{\mathbb{T}_{\mathfrak{c}}}(t_j) \end{matrix} \right. , \quad i \neq j \in \mathbb{R}^+, \quad t \in \mathbb{T} = (-\infty, +\infty)$$

where

$$\mu_{\mathbb{T}_{\mathfrak{b}}}(t_i) = \mu_{\mathbb{T}_{\mathfrak{b}}}(t_j) \quad \text{and} \quad \mu_{\mathbb{T}_{\mathfrak{c}}}(t_i) = \mu_{\mathbb{T}_{\mathfrak{c}}}(t_j) \quad \text{if } i = j \Rightarrow t_i = t_j$$

Axiom 3.3.3.1.1 simply affirms that near future is more valuable than the distant future for the same unit of benefit. Similarly, for any unit cost at the same time point the society assesses the distant future to be more valuable than the near future and hence prefer to defer the cost. Alternatively, the society prefers current outcome if it is conceived as a benefit while the same society prefers future outcomes if it is conceived as cost, where benefits and costs are measured in the same unit scale. Transformation-decision time preference Axiom 3.3.3.1.1 may be complemented with Axiom 3.3.3.1.2.

Axiom 3.3.3.1.2 *Comparability of Benefit and Cost Times*

For any given time, $t \in \mathbb{T}$, i and $j \in \mathbb{R}^+$ (the reals) and for the same unit value of cost and benefit, one of the following must hold:

$$\left\{ \begin{matrix} \text{a)} \\ \text{b)}\ \mu_{\mathbb{T}_{\mathfrak{b}}}(t_j) \\ \text{c)} \end{matrix} \right. \left\{ \begin{matrix} > \\ = \\ < \end{matrix} \right\} \mu_{\mathbb{T}_{\mathfrak{c}}}(t_i), \text{ if } i \left\{ \begin{matrix} > \\ = \\ < \end{matrix} \right\} j, \text{ given that } \left\{ \begin{matrix} \dot{\mu}_{\mathbb{T}_{\mathfrak{b}}}(\cdot) \leq 0 \\ \dot{\mu}_{\mathbb{T}_{\mathfrak{c}}}(\cdot) \geq 0 \end{matrix} \right. ,$$

where $d\mu/dt = \dot{\mu}_{\mathbb{T}}$ is the slope.

For all admissible membership functions, Axiom (3.3.3.1.2) presents comparability conditions where the grade values of benefit time and cost time can be compared given the same unit of measurement of costs and benefits. Axioms 3.3.3.1.1 and 3.3.3.1.2 may be strengthened. Let the time set, $\mathbb{T}$, be induced by a fuzzy partition into near future time set, $\mathbb{N}$ and distant future time set, $\mathbb{D}$, where $\mathbb{N} \bigcup \mathbb{D} = \mathbb{T}$ and $\mathbb{N} \bigcap \mathbb{D} = \emptyset$ and where $\tilde{\mathbb{T}}$ is a general fuzzy time set (Note: a fuzzy time set, $\tilde{\mathbb{T}}$, is different from ordinary time set, $\mathbb{T}$).

Axiom 3.3.3.1.3 *Comparability of Cost and Benefit Time Sets in Near-distance Time Sets*

For any given $t \in \mathbb{T}$ and any cost and benefit sets $\mathbb{C}$ and $\mathbb{B}$ respectively

$$\begin{matrix} \text{a)} \\ \text{b)}\,\mu_{\mathbb{T}_b}(t) \\ \text{c)} \end{matrix} \begin{Bmatrix} \succ \\ \prec \\ \sim \end{Bmatrix} \mu_{\mathbb{T}_c}(t) \quad \text{if} \begin{Bmatrix} t \in \mathbb{N} \subset \mathbb{T} \\ t \in \mathbb{D} \subset \mathbb{T} \\ t \in (\mathbb{D} \bigcap \mathbb{N}) \subset \mathbb{T} \end{Bmatrix}$$

Axioms 3.3.3.1.1, 3.3.3.1.2 and 3.3.3.1.3 demonstrate an evaluative conflict of degree of intensity of preference between fuzzy cost time and fuzzy benefit time. The conflict reveals itself as decision agents prefer benefit over cost and will choose decision action with benefit if the cost can be avoided but this is behavioral violation of the *asantrofi-anoma* principle where every benefit has a cost-support and vice versa in all decision-choice processes that demand appeals to cost-benefit rationality. This evaluative conflict may be reconciled by formulating the conflict as a fuzzy decision problem. The solution of the fuzzy decision problem provides channels for ranking the transformation-decisions in the dynamics of actual-potential polarity under the cost-benefit rationality and the principle of *asantrofi-anoma* over the space of time-points of fuzzy inter-temporal preferences. The membership function of the fuzzy benefit time set may be viewed as an objective function while the membership function of fuzzy cost time set acts as a constraint on the valuation of the usefulness of transformation-decision of the actual variety with a replacement of a potential variety.

In other words, the valuation of social action on the transformation decision in an environment of fuzzy-stochastic uncertainty and risk may be undertaken on the basis of pure preferences of transformation-decision time involving relative assessments of fuzzy costs and fuzzy benefits. This evaluation of whether a particular time is appropriate for social transformation may then be specified as a fuzzy decision problem, Δ where the transformation-decision involves the destruction of the existing actual $\mathfrak{a} \in \mathfrak{A}$ and the replacement of a potential $v \in \mathfrak{V} \subset \mathfrak{P}$. The fuzzy decision problem may be written as:

$$\Delta_{\mathfrak{V}} = \left\{ \left(t, \mu_{\Delta_{\mathfrak{V}}}(t)\right) \middle| \mu_{\Delta_{\mathfrak{V}}}(t) = \left(\mu_{\tilde{\mathbb{T}}_b}(t) \wedge \mu_{\tilde{\mathbb{T}}_c}(t)\right), t \in \mathbb{T} \right\} \tag{3.3.3.1.1}$$

where $\wedge$ is a min-operator. A fuzzy decision problem differs from the classical decision problem in terms of the information environment that the problems are defined as well as the algorithms to abstract the corresponding solutions. For discussions and mathematics on fuzzy decision problems make reference to [R2] [R2.13], [R4], [R4.14], [R5], [R55.16], [R5.26], [R5.34], [R6], [R6.6], [R6.8].

The fuzzy transformation-decision problem of appropriate time of the social actual is to find the value of relative cost-benefit information that will optimize the stream of net benefits given the cost-benefit flows associated with all future time points of a potential variety replacement of the current actual where the potential varieties are in the feasible set $v \in \mathfrak{V} \subset \mathfrak{P}$. Alternatively, the society is to value time so as to maximize the stream of benefit flows subject to the condition that each time benefit must be time-cost supported under the *Asantrofi-anoma principle* and solution to the *Asantrofi-anoma problem* of optimal choice. The time valuation on the basis of the degree of societal preference may be obtained by solving the fuzzy optimization problem of the form

$$\underset{t\in\mathbb{T}}{\mathrm{opt}}\,\mu_{\Delta_{\mathfrak{v}}}(t) = \sup_{t\in\mathbb{T}}\left[\mu_{\tilde{\mathbb{T}}_{\mathfrak{b}}}(t) \wedge \mu_{\tilde{\mathbb{T}}_{\mathfrak{c}}}(t)\right] \tag{3.3.3.1.2}$$

The solution to the fuzzy optimization problem may be obtained by using the method of fuzzy mathematical programming [R6], [R6.3], [R6.6], [R6.9], [R6.11]. Thus, we may state Eq. (3.3.3.1.2) as an equivalence theorem.

Theorem 3.3.3.1.1 *The fuzzy optimization problem* $\mathrm{opt}\,\mu_{\Delta_{\mathfrak{v}}}(t)$, *is equivalent to solving the problem*

$$\underset{t\in\mathbb{T}}{\mathrm{opt}}\,\mu_{\Delta_{\mathfrak{v}}}(t) = \begin{cases} \inf\limits_{t\in\mathbb{T}^*} \mu_{\mathbb{T}_{\mathfrak{b}}}(t) \\ s.t \quad \mathbb{T}^* = \left\{t\in\mathbb{T} \middle| \left[\mu_{\mathbb{T}_{\mathfrak{b}}}(t) - \mu_{\mathbb{T}_{\mathfrak{c}}}(t)\right] \geq 0\right\} \end{cases}$$

Corollary 3.3.3.1.1

$$\underset{t\in\mathbb{T}}{\mathrm{opt}}\,\mu_{\Delta_{\mathfrak{v}}}(t) = \begin{cases} \sup\limits_{t\in\mathbb{T}^*} \mu_{\mathbb{T}_{\mathfrak{c}}}(t) \\ s.t \quad \mathbb{T}^* = \left\{t\in\mathbb{T} \middle| \left[\mu_{\mathbb{T}_{\mathfrak{b}}}(t) - \mu_{\mathbb{T}_{\mathfrak{c}}}(t)\right] \geq 0\right\}. \end{cases}$$

The proofs of these theorems are equivalent to those that have been given in [R5.41], [R6.3]. For recent advances in fuzzy optimization see [R6.6]. Suppose that $t = t^*$ solves the problem defined by Eq. (3.3.3.1.2) then the optimal transformation-decision time will involve the situation where the relative cost-benefit information is optimal. The individual optimal transformation-decision relative cost-benefit time $\mathfrak{r}*$, that reconciles benefit time and cost time valuations is

$$\mathfrak{r}_{\mathrm{v}}^* = (\Xi_{\mathfrak{a}}/\Xi_{\mathfrak{p}})^* = \left(\left(\frac{\mathfrak{C}_{\mathfrak{p}}}{\mathfrak{C}_{\mathfrak{a}}}\right)\left(\frac{\mathfrak{B}_{\mathfrak{a}}}{\mathfrak{B}_{\mathfrak{p}}}\right)\right)^* = \mu_{\mathbb{T}_{\mathfrak{b}}}(t^*) = \mu_{\mathbb{T}_{\mathfrak{c}}}(t^*), \forall \mathrm{v} \in \mathfrak{V} \text{ and } \mathfrak{a} \in \mathfrak{A} \tag{3.3.3.1.3}$$

Equation (3.3.3.1.3) spins a distribution $\mathfrak{R}_{\mathfrak{V}}^*$ over $\mathrm{v} \in \mathfrak{V} \subset \mathfrak{P}$ in that every $\mathrm{v} \in \mathfrak{V} \subset \mathfrak{P}$ has a corresponding optimal relative cost-benefit time $\mathfrak{r}_{\mathrm{v}}^*$ that can be put in decreasing or increasing social rank order and the maximum selected value to provide $\mathfrak{r}_{\mathrm{v}*}^* \in \mathfrak{R}_{\mathfrak{V}}^*$.

The fuzzy transformation-decision problem defined with its solution is represented in a geometric form in Fig. 3.3.

Theorem 3.3.3.1.2 *If the fuzzy decision is convex then there exist* $t^* \in \mathbb{T}$ *such that* $\mu_{\mathbb{T}_b}(t^*) = \mu_{\mathbb{T}_c}(t^*)$ *and* $\mu_\Delta(t^*)$ *is optimal and unique.*

The proof of this theorem may be constructed by using the regularity conditions of fuzzy preferences, Theorem (3.3.3.1.1) and conditions of fuzzy convex decision [R2.22], [R2.36], [R5.41], [R5.52], [R5.63]. The appropriate time of a social

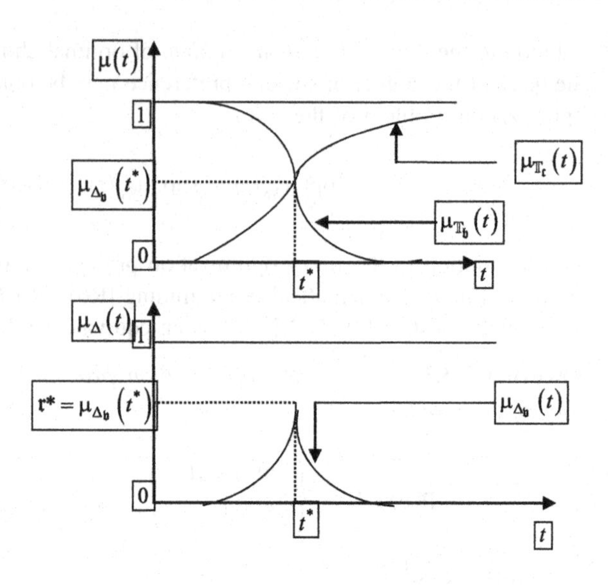

Fig. 3.3 The geometric solution of the optimal transformation-decision time in the dynamics of social actual-potential polarity and an information process

transformation of an existing actual to a new actual by the social decision process provides equal social assessed weights for cost time and benefit time at the same time point. At this optimal transformation-decision time, the society is willing to accept the cost of social transformation of the existing actual in terms of opportunity of benefit forgone as well as willing to pay for the cost of actualizing the new potential for the benefit that accompanies the new actual from the potential. This is the fuzzy equilibrium transformation-decision of cost-benefit time $\mathfrak{r}^*_{v*} \in \mathfrak{R}^*_{\mathfrak{V}}$ and the corresponding social variety $v^* \in \mathfrak{V}$ which is the optimal actualized social variety at optimal cost-benefit time and in replacement of the current actual $v_a \in \mathfrak{A}$ which fades into the potential $v_a \in \mathfrak{P} = \mathfrak{U}$. The actualized potential variety brings in a new cost-benefit relative information flow (info-flow) of the form $\mathfrak{S}_{v^* \in \mathfrak{V}} = (\#\mathbb{Z}_{\mathfrak{B}}/\#\mathbb{Z}_{\mathfrak{C}})_{v^* \in \mathfrak{V}}, \forall t \in \mathbb{T}$. The disappearance of the actual variety $a \in \mathfrak{A}, \forall t \in \mathbb{T}$ leaves its blueprint of information $\mathfrak{S}_{a \in \mathfrak{A}} = (\#\mathbb{Z}_{\mathfrak{B}}/\#\mathbb{Z}_{\mathfrak{C}})_{a \in \mathfrak{A}}, \forall t \in \mathbb{T}$ behind as a part of the available info-stock in the space of the actual which is composed of the sub-spaces of known and unknown actual.

3.3.3.2 Time, Decision, Transformation and the Info-Process

In any social setup, social transformation is a collective decision, where such a decision at any transformation-decision time point is locked in individual conflicts over the collective decision space within the epistemological space in which cognitive agents operate. The individual conflicts find expressions in preferences of information as revealed by time, costs and benefits. This is because any social transformation decision has two interrelated actions under the principles of opposites in duality expressed over social polarities in terms of information generation. At a level of generality, a social transformation is a revolution against the existing actual, and a destruction of the social actual variety. It is also an important contest for other potential varieties that may be actualized in place of the existing actual variety. The individual conflicts first reveal themselves over the actual in terms of individual relative cost-benefit assessments. Here, some individuals have preference for the existing social actual variety, some other individuals have preference for the destruction of the existing social variety $\nu_{\mathfrak{a}_0^*} \in \mathfrak{A}$ and for a transformation into a new social variety $\nu \in \mathfrak{V}$, while some individuals may be indifferent.

The transformation decision-choice problem comes with a numbers of complications that create epistemic complexities. The presence of the existence and non-existence of a variety is the cost-benefit conditions of a *with-and-without* social decision problem which is a collective one that may be undertaken by either a democratic or non-democratic process. The nature of this problem and its solution under democratic social decision-choice system and its implications for non-democratic action have been discussed extensively in [R13.8], [R13.9]. A new problem arises from the solution to the with-and-without problem. The new problem reveals itself in terms of the cost-benefit timing of the transformation-substitution process of the social variety of the transformation which requires the knowledge of the *necessary conditions* and the creation of the *sufficient conditions* to bring about a result. The decision-choice action of an optimal time is also a collective decision-choice action that may be democratic or non-democratic under the value $\mathfrak{r}_{\nu*}^* \in \mathfrak{R}_{\mathfrak{V}}^*$. Given that the democratic collective decision-choice action is in the favor of the destruction of the existing social variety and the creation of a new social variety with a choice of an optimal time a new problem arises. This problem centers around the appropriate potential $\nu \in \mathfrak{V}$ to be actualized into a new actual $\nu_{\mathfrak{a}_n^*} \in \mathfrak{A}$ to replace the old actual social variety $\nu_{\mathfrak{a}_0^*} \in \mathfrak{A}$ in the transformation-substitution process. The decision-choice action on the appropriate potential is also a collective decision-choice problem that may be resolved by democratic or non-democratic action.

Each problem-solution process generates two information structures where the information $\mathbb{Z}_{\nu_{\mathfrak{a}_0^*}}$ on the old actual variety $\nu_{\mathfrak{a}_0^*} \in \mathfrak{A}$ remains as part of the info-stock and the information structure $\mathbb{Z}_{\nu_{\mathfrak{a}_n^*}}$ of the new actual $\nu_{\mathfrak{a}_n^*} \in \mathfrak{A}$ is a flow that goes to update the info-stock. At any moment of transformation-decision time point $t \in \mathbb{T}$, there is the actualized info-stock $\mathbb{Z}_{\mathfrak{A}_t^*}^{\nu}$ where $\mathbb{Z}_{\mathfrak{A}_t^*}^{\nu} = \bigcup_{(t=-\infty)}^{(t>-\infty)} \mathbb{Z}_{\mathfrak{A}_t^*}^{\nu}$ and $\mathbb{Z}_{\mathfrak{A}_{-\infty}^*}^{\nu} = \bigcap_{(t=-\infty)}^{(t>-\infty)} \mathbb{Z}_{\mathfrak{A}_t^*}^{\nu}$ is the info-stock at the beginning regarding the variety $\nu \in \mathfrak{V}$. The info-stock $\mathbb{Z}_{\mathfrak{A}_t^*}^{\nu} = \bigcup_{(t=-\infty)}^{(t>-\infty)} \mathbb{Z}_{\mathfrak{A}_t^*}^{\nu}$ is a derived characteristic-signal disposition for the primary characteristic-signal disposition $\mathbb{Z}_{\mathfrak{A}_{-\infty}^*}^{\nu} = \bigcap_{(t=-\infty)}^{(t>-\infty)} \mathbb{Z}_{\mathfrak{A}_t^*}^{\nu}$. In the info-time process, the total actualized derived info-stock is $\mathbb{Z}_{\mathfrak{A}_t^*}^{\mathfrak{V}} = \bigcup_{\nu \in \mathfrak{V}} \bigcup_{(t=-\infty)}^{(t>-\infty)} \mathbb{Z}_{\mathfrak{A}_t^*}^{\nu}$ and the primary info-stock is $\mathbb{Z}_{\mathfrak{A}_{-\infty}^*}^{\mathfrak{V}} = \bigcap_{\nu \in \mathfrak{V}} \bigcap_{(t=-\infty)}^{(t>-\infty)} \mathbb{Z}_{\mathfrak{A}_t^*}^{\nu}$.

The info-process is continual activities under transformation-substitution principle of variety destruction and variety creation without the destruction of information on old varieties but with continual flow of information on new actualized varieties as addition to the info-stock. Thus the info-process is a continual destruction of old varieties without destruction of the corresponding information and the creation of new varieties and new flow of information. The important idea about this process is that the size of the set of the initial varieties $\mathbb{V}_{-\infty}$ may expand or shrink as seen in time (t) to obtain $\mathbb{V}_t$ where $\#\mathbb{V}_t < \#\mathbb{V}_{-\infty}$ in the case of variety shrinkage and $\#\mathbb{V}_t > \#\mathbb{V}_{\infty}$ in the case of variety expansion or no change in which case $\#\mathbb{V}_t = \#\mathbb{V}_{-\infty}$. It must be understood that the shrinkage, expansion or no-change does not mean that the same varieties are existing in the different time points. The quantitative varieties may be the same but the qualitative varieties are constantly on transformation. The information on the destroyed varieties is in the info-stock while the information on the new varieties adds to the existing info-stock.

Chapter 4
The Theory of Info-dynamics: An Introduction to Its Conceptual Frame

It is useful now to provide an epistemic framework on which the theory of info-dynamics is being developed. This theory is a follow up of the theory of info-statics. It is developed as a unified foundational theory of informing through acquaintances, knowing and learning through epistemic processes which take the forms of teaching and educating through transmission and communication processes of conditions on transformation of varieties and categorial varieties. The transformations of varieties and categorial varieties take through decision-choice processes and the mimicking of nature to satisfy the needs of engineering processes.

4.1 Some General Reflections on Epistemic Directions

Chapter 1 of this monograph begins with a reflection and critique on the traditions of information theory as a prelude to the theory of info-dynamics. The essential criticism is seen from the viewpoint that the theory of information is made up of two sub-theories that may be interdependent or not. The first sub-theory of relevance is the *theory of information contents* (info-content). The second sub-theory is the *theory of communication* of the contents among ontological elements. The information contents allow the establishment of varieties, categorial varieties and categories through distinctions, differences and commonness. The information communication is the sharing of the contents through source-destination processes while the information transmission is the revelation of the contents which allows the establishment of a framework of an awareness through the acquaintances with varieties and categorial varieties. The contents of information about varieties and categorial varieties are established by the *characteristic dispositions*. The transmissions and communications of the contents are done through the *signal dispositions* of the characteristic dispositions. The theory of info-contents, the theory of info-transmission and the theory of info-communication constitute the *theory of info-statics* which is essentially about the definitions of information to establish

K.K. Dompere, *The Theory of Info-Dynamics: Rational Foundations of Information-Knowledge Dynamics*, Studies in Systems, Decision and Control 114, https://doi.org/10.1007/978-3-319-63853-9_4

contents, transmission of information contents between ontological objects on one hand and epistemological objects on the other hand and the communication of the information contents among classes of epistemological objects.

The theory of info-statics helps to establish conditions of informing, knowing, learning and teaching and their effects on *transformation-decision systems* of varieties and categorial varieties leading to the information production. It is useful to keep in mind that every decision is about affirmation of existing variety or a change of existing variety in the actual space $\mathfrak{A}$ and actualization of a new variety from the potential space $\mathfrak{P} = \mathfrak{U}$. The process is the dynamics of actual-potential polarity. There is no distinction between the potential space and the possibility space and there is no existence of probability space in ontological transformations. Every potential variety is also a possible variety. The transformations are either from the actual to the potential or from the potential to the actual without uncertainties. The distinction between the possibility space and the potential space and the introduction of probability space are relevant in epistemological activities of all kinds. Here, epistemological transformations and activities go from the potential space $\mathfrak{U}$ through the possibility space $\mathfrak{P}$ and through the probability space $\mathfrak{B}$ to reach the space of the actual $\mathfrak{A}$. The set of instruments of internal transformations of varieties is the set of potential and actual dualities with relational continua and unity under the general principle of opposites. The understanding of the information production process and its behavior through the dynamics of the actual-potential polarity is studied under the *theory of info-dynamics*. The epistemic differences and similarities between the subject matters of info-statics and info-dynamics and their corresponding theories have been taken up in Chap. 2 of this monograph. It is argued that the theory of info-statics establishes the notion that information is a property of matter in varieties and categorial varieties and the transmissions and communications of the contents of the information to establish an awareness of varieties and categorial varieties are energy processes. The theory of info-dynamics establishes the framework of continual transformation of the varieties and categorial varieties through transformation-decision activities over both the ontological and epistemological spaces where such transformations are energy processes.

In the previous monographs, the concepts of necessity and *freedom* were introduced as an interdependent universal conditions where the necessity-freedom connection is established in relation to the possible and the probable with further connection to necessary-sufficient conditions in the socio-natural transformation space. The concepts of possible and probable do not arise in the dynamics of the ontological varieties. The concept of possible and probable are elements in the dynamics of epistemological varieties and cognitive activities over the epistemological space. When considerations are given to the behavior of cognitive agents, a distinction must be established between the potential and the possible and hence the potential space $\mathfrak{U} = \Omega$ may not be viewed as the same as the space of possibilities. Over the epistemological space necessity is related to the *possibility space* that generates the *necessary conditions* of the epistemological transformations of varieties. The explanatory structure of the necessary conditions is provided by the theory of categorial conversion [R17.15]. The freedom is related to the *probability space* $\mathfrak{B}$ that generates the *sufficient conditions* of transformation. The explanatory structure of the sufficient conditions is provided by the theory of Philosophical Consciencism [R17.16]. In other words, necessity and freedom are foundations of the development of the *general theory of transformation* that forms the essential framework in the development of the theory of info-dynamics. In this monograph, the theory of info-dynamics is developed to link the necessity-freedom connection of transformations of universal existence to information stock-flow processes on varieties and categorial varieties.

The general theory of transformation is, thus, split into two inter-dependent sub-theories. One sub-theory reveals the working mechanism of the inner necessities of varieties that present the set of necessary conditions which makes it possible for self-motion and self-transformation. This is the *theory of Categorial conversion* that has been summarized and fully developed in the monograph with the same title [R17.15]. The other sub-theory reveals the working mechanism of the inner sufficiency of command and control of the necessary conditions of varieties and categorial varieties that present a set of sufficient conditions which makes it probable to effect self-motion and self-transformation. This is the theory of Philosophical Consciencism that has been summarized and fully developed under the same title in [R17.16]. At the level of the general theory of information, all optimization problems and corresponding solutions are developed to reconcile the conditions of necessity and freedom, where freedom is optimize subject to necessity within the cost-benefit duality. The conditions of necessity may be viewed as *cost conditions* of the socio-natural transformation decisions while the conditions of freedom may be viewed as *benefit conditions* of socio-natural transformation decisions where optimal freedom is sought.

The general theory of transformation is about destruction and creation of varieties and categorial varieties in a movement from the potential to the actual in a creative process, and from the actual to the potential in destructive process under the constructionism-reductionism principle within the dynamics of actual-potential polarities. Each variety has an inner necessity that defines the necessary conditions for the inner dynamics and the generation of the requirements for change. The transformation of varieties under these requirements for change is brought into fruition though the development of the sufficient conditions for transformation of varieties and categorial varieties. The epistemic framework is that both the inner necessity and categorial conversion are related to the necessary conditions of transformation, and are defined in the *possibility space* $\mathfrak{P}$ which provides the internal necessary conditions for self-motion and self-transformations of varieties. The freedom and the internal Philosophical Consciencism relate to the creation of the sufficient conditions through the management of the control-command decision systems of the inner necessary conditions. For the transformation of social varieties the freedom and the internal Philosophical Consciencism are defined in the *probability space* $\mathfrak{B}$. The possibility space constrains the size of the probability space as well as the probable outcomes that may be defined in the probability space.

It is here that agents of transformation of social varieties have freedom of action which is limited by the information-knowledge process in the transition from the possibility space $\mathfrak{P}$ to the probability space $\mathfrak{B}$ and to the space of the actual $\mathfrak{A}$. Necessity in transformation-decisions of varieties and categorial varieties is a constraint on freedom of *intentionality* to undertake preferred transformation-decisions to produce new varieties and new information flow. It is just like the categorial-conversion process where the varieties over the possibility space constrain the changes of the variety outcomes in the probability space to engender the actualized outcomes of new varieties and hence flows of information in the space of the actuals. Freedom in transformation-decisions of varieties is the objective to optimize transformation-decision-choice intentionality to create an efficient management of command-control processes to produce new preferred varieties and hence new information on the varieties. It is just like the process of Philosophical Consciencism where there are the developments of appropriate strategies in the probability space to destroy the existing actual varieties and actualize an element or some elements in the feasible varieties in the possibility space through the dynamics of the actual-potential polarity under the game strategy of the principle of opposites. The study of game strategies in the behavior of the opposites is essential to the understanding of the information stock-flow conditions through the destruction-creation process within in the cost-benefit time. Analytically, necessity relates to information that cognitive agents hold over the possibility space regarding possible varieties while freedom relates to knowledge-decision choice process to effect transformations. Knowledge-decision-choice process is thus optimized subject to an information constraint over the epistemological space of social transformations under the general matter-energy process.

The whole matter-energy process to produce stock-flow information involves three elements of quantity, quality and time on the basis of which the theory of info-dynamics is developed and understood. Time is neutral irrespective of how it is measured, however, an appropriate concept of time is useful in the understanding of the content production and transmission of information. The appropriate concept of time and the development of its structural behavior are provided in Chap. 3 to allow a distinction of a time point and period to be specified Given the concept of time, the theory of info-dynamics involves three types of dynamic systems. One type involves the quality-time phenomena holding quantity constant. This is called the inter-categorial dynamics involving transformations of varieties in qualitative differences and similarities. The second type involves the quantity-time phenomena holding quality constant. This is called the intra-categorial dynamics involving transformations of varieties in quantitative differences and similarities. The intra-categorial dynamics includes the space-time phenomena. There is a third one which involves a simultaneity of quality-quantity-time phenomena. This may be called the complete-categorial dynamics where quantity and quality of varieties simultaneously change with time. These three types present complexities in the matter-energy systems dynamics. To deal with these complexities, different concepts of time are discussed in Chap. 3.

4.1.1 *Differences and Similarities of Varieties in the Quantity-Time and Quality-Time Spaces*

The general concept of variety relates to quantity and quality phenomena. The concept breaks done into two types of inter-categorial and intra-categorial varieties. These concepts of inter-categorial variety and intra-categorial variety are easier to conceptualize in info-statics when motion is not contemplated. When transformation dynamics are contemplated, there are two classes of motion that have been discussed relative to information production and communication as info-dynamics. At the level of info-statics, the important concepts of variety are *inter-categorial (qualitative) varieties* such as different types of animals or objects in the universe producing qualitative categories and qualitative information in the universal system with qualitative and intra-categorial (quantitative) varieties. There is also the *intra-categorial (qualitative) varieties* such as the sizes of dogs producing quantitative categories within a given qualitative category and quantitative information.

The concept of variety acquires some important complexities especially in the quantity-time space such as space-time phenomena. Here, varieties are defined by an object with the constant quality in different positions in either linear or non-linear space. The characteristic-signal disposition is conceptualized as an object-time-position phenomenon that generates *qualitative information* in the intra-categorial conversion which is basically a conversion of position-variety in the space relative to time. The characteristic disposition is defined as position-time property such that given an object ω and ϕ as the corresponding phenomenon defining an inter-categorial variety of the form $v = (\omega, \phi) \in (\Omega \otimes \Phi)$, an intra-categorial variety may be defined as a transformation on inter-categorial variety $v \in \mathbb{V}$ written as $\hat{v} = \mathfrak{T}_v(\cdot)$. Let $\mathfrak{P}$ be the set of positions with a generic element $\mathfrak{p} \in \mathfrak{P}$, $\mathfrak{V}$ the set of velocities with a generic element $\mathfrak{v} \in \mathfrak{V}$, $\mathfrak{A}$ the set of accelerations with generic element $\mathfrak{a} \in \mathfrak{A}$, the set of sizes $\mathfrak{S}$ with generic element $\mathfrak{s} \in \mathfrak{S}$, a set of distances $\mathfrak{Y}$ with a generic elements $\mathfrak{y} \in \mathfrak{Y}$ an intra-categorial variety may be defines as $\hat{v} = \mathfrak{T}_v(\mathfrak{p}, \mathfrak{v}, \mathfrak{a}, \mathfrak{s}, \mathfrak{y}, t)$ where $\mathfrak{p} \in \mathfrak{P}, \mathfrak{v} \in \mathfrak{V}, \mathfrak{a} \in \mathfrak{A}, \mathfrak{y} \in \mathfrak{Y}, \mathfrak{s} \in \mathfrak{S}$ and $t \in \mathbb{T}$. The state of the new position-time variety is given as $\mathfrak{r}(t) = (\mathfrak{p}, \mathfrak{v}, \mathfrak{a}, \mathfrak{s}, \mathfrak{y})$ which provides the quantitative info-dynamics, where the size $\mathfrak{s}$, distance $\mathfrak{y}$ and position $\mathfrak{p}$ may be interchangeable, depending on the nature of inter-categorial variety and the quantitative disposition of the distribution of quantitative-signal dispositions that presents information distribution on intra-categorial varieties.

4.1.2 *Differences and Similarities in Information Representation of Inter-categorial and Intra-categorial Transformation Dynamics*

Let us recall that corresponding to any variety $\mathrm{v}_k \in \mathbb{V}$ there is a corresponding characteristic-signal disposition of the form $\mathbb{Z}_k = (\mathbb{X}_k \otimes \mathbb{S}_k) = \{z_j = (x_j, s_j) | j \in \mathbb{J}_k \subset \mathbb{J}^{\infty}, k \in \mathbb{L}^{\infty}\}$ that allows the identity of the variety to be distinguished. Let $\mathrm{v}_k \in \mathbb{V}$ be the inter-categorial varieties with a corresponding characteristic-signal disposition as $\mathbb{Z}_k = (\mathbb{X}_k \otimes \mathbb{S}_k), \forall k \in \mathbb{L}^{\infty}$. The intra-categorial variety may then be specified as $\widehat{\mathrm{v}} \in \widehat{\mathbb{V}}$ with characteristic-signal disposition specified as $\widehat{\mathbb{Z}}_k = \widehat{(\mathbb{X}_k \otimes \mathbb{S}_k)} = (\widehat{\mathbb{X}}_k \otimes \widehat{\mathbb{S}}_k)$. The theory of info-statics is the study of varieties in time and the stock of information about these varieties that are presented as the inter-categorial varieties and intra-categorial varieties at that point in time. It is thus a study of the explanatory and the prescriptive information at a time point. The theory of info-dynamics is the study of transformations of varieties over time and about the flows of information that are generated by the transformations of varieties to update the info-stock.

In this epistemic view, the general information theory is the study of ontological and epistemological varieties $\mathbb{W}$ which is composed of inter-categorial varieties $\mathbb{V}$ with information $\mathbb{Z}$ and intra-categorial varieties $\widehat{\mathbb{V}}$ with information $\widehat{\mathbb{Z}}$ where the total varieties is such that $\mathbb{W} = \mathbb{V} \bigcup \widehat{\mathbb{V}}$ with information $\mathbb{Z}_{\Omega} = \mathbb{Z} \bigcup \widehat{\mathbb{Z}}$.

This framework simply means that the general set of varieties is composed of a subset of qualitative varieties $\mathbb{V}$ and a subset of quantitative varieties $\widehat{\mathbb{V}}$. The set of inter-categorial varieties $\mathbb{V}$ is composed of the sub-set of inter-categorial actual varieties $\mathbb{V}_{\mathfrak{a}}$ with corresponding information support $\mathbb{Z}_{\mathfrak{a}}$ and a sub-set of inter-categorial potential varieties $\mathbb{V}_{\mathfrak{p}}$ with corresponding information support $\mathbb{Z}_{\mathfrak{p}}$ such that $(\mathbb{V} = \mathbb{V}_{\mathfrak{a}} \bigcup \mathbb{V}_{\mathfrak{p}})$ and $(\mathbb{Z} = \mathbb{Z}_{\mathfrak{a}} \bigcup \mathbb{Z}_{\mathfrak{p}})$. The set of intra-categorial varieties $\widehat{\mathbb{V}}$ is also composed of a set of actual intra-categorial varieties $\widehat{\mathbb{V}}_{\mathfrak{a}}$ and a set of potential intra-categorial varieties $\widehat{\mathbb{V}}_{\mathfrak{p}}$ such that $\left(\widehat{\mathbb{V}} = \widehat{\mathbb{V}}_{\mathfrak{a}} \bigcup \widehat{\mathbb{V}}_{\mathfrak{p}}\right)$ with an information support $\left(\widehat{\mathbb{Z}} = \widehat{\mathbb{Z}}_{\mathfrak{a}} \bigcup \widehat{\mathbb{Z}}_{\mathfrak{p}}\right)$ that allows identities to be revealed for distinctions and differences of varieties and categorial varieties. At the level of ontology, the potential space is the same as the possibility space without uncertainties and risks. At the level of epistemology, however, there is a cognitive separation between the potential space and the possibility space induced by cognitive limitations of cognitive agents where the possibility space is conceived as a sub-space of the potential space for knowing, learning and teaching.

It may be kept in mind that the space of the actual $\mathfrak{A}$ is the space of inter-categorial and intra-categorial varieties which is composed of a set of claimed known varieties $\mathfrak{A}^{\mathfrak{n}}$ and a set of unknown varieties $\mathfrak{A}^{\mathfrak{u}}$ such that in general and at any time point $t \in \mathbb{T}$ it is the case that $(\mathfrak{A} = \mathfrak{A}^{\mathfrak{n}} \bigcup \mathfrak{A}^{\mathfrak{u}})$. It is also the case that

$\#\mathfrak{A}^{\mathfrak{n}} \lll \#\mathfrak{A}^{\mathfrak{u}}$ where the space of the known actual varieties $\mathfrak{A}^{\mathfrak{n}}$ expands through the knowledge-production process with the supporting information structure, and the space of unknown actual varieties $\mathfrak{A}^{\mathfrak{u}}$ and the supporting information $\mathbb{Z}^{\mathfrak{n}}$ expand through the transformation process and the dynamics of actual-potential polarities. The identity of any variety ν has implied information, data, fact and evidence, the definitional structures of which have been discussed in [R17.17]. To construct a theory of general transformation of varieties and its relationship to information behavior, one must introduce time into the definitional blocks of the info-statics to allow the examination of each time-point conditions. The epistemic process is to transform the static conditions of information content and communication to dynamic conditions of information content and communication in a manner that allows a comparison of different time-point behavior of the same object and phenomenon $(\omega, \phi) \in (\Omega \otimes \Phi)$ with a corresponding information structure $(\mathbb{Z} = \mathbb{X} \otimes \mathbb{S})$ defined in terms of characteristic-signal deposition. The comparison is about the changing time-point variety of the same element in the space-time process.

It is always analytically helpful to carry in mind that the general information system to which an epistemic construct is being carried on is made up of the object set Ω with generic element $\omega \in \Omega$, a set of phenomena Φ with a generic element $\phi \in \Phi$, a set of characteristics $\mathbb{X}$ with a generic element $x \in \mathbb{X}$, a set of signals $\mathbb{S}$ with a generis element $s \in \mathbb{S}$. The object set and the set of phenomena create a variety space $\mathbb{V}$ with a generic element $\nu \in \mathbb{V}$ such that each of the variety $\nu \in \mathbb{V}$ is defined by an object $\omega \in \Omega$ and phenomenon $\phi \in \Phi$ of the form $(\omega, \phi) \in (\Omega \otimes \Phi)$. The corresponding identity of each variety is revealed by the general information structure of the form $(\mathbb{Z}_\Omega = \mathbb{X}_\Omega \otimes \mathbb{S}_\Omega)$ and a specific information on the varieties of the form $(\mathbb{Z}_\nu = \mathbb{X}_\nu \otimes \mathbb{S}_\nu), \forall \nu \in \mathbb{V}$. These concepts and symbolic representations hold at each pure time point. Together these concepts and their representations define a complex system which points to the idea that the *theory of knowledge* is nothing more that the *theory of variety search* where the naming of varieties is through *methodological nominalism* and the discovery of varieties is through the methodological *constructionism-reductionism duality* with relational continuum and unity. The complex system is such that the theory of socio-natural change is nothing more than the study of information on variety transformations.

The general structure of the study of variety transformations in all arears of ontological existence in the epistemological space is here referred to as the theory of info-dynamics. All transformations have a set of *necessary conditions* and a set of *sufficient conditions*. The necessary conditions exhibit external indications for change and the study of which is dealt with in the *Theory of Categorial Conversion* [R17.15]. The sufficient conditions exhibit internal forces to bring about a change and the study of which is dealt with in the *Theory of Philosophical Consciencism* [R17.16]. The general change of varieties at any pure time point and period is information-decision-choice interactive processes, where the set of the necessary conditions establishes a *necessity* and the set of sufficient conditions establishes a *freedom* in a dynamic transformation-decision domain. The necessity is a constraint on freedom just as the set of conditions of categorial conversion is a constrained on the set of conditions of Philosophical Consciencism. At the epistemological space, the transformation of social actual-potential polarity is a maximization of freedom in the decision-choice space subject to a necessity to provide the best decision-choice variety among a set of varieties under the control of cognitive agents. It is this recognition of freedom-necessity relational structure, that the understanding of the necessity finds meaning through the epistemic framework of the theory of categorial conversion, where necessity relates to transformation cost conditions. It is the same recognition of the freedom-necessity relational structure, that the understanding of freedom finds meaning through the epistemic framework of the theory of Philosophical Consciencism, where freedom relates to transformation benefit conditions. In this necessity-freedom structure, goals and intentionality interact with decision-choice actions to destroy existing varieties and create new varieties in their substitute with continual info-dynamic processes. The relational structure of information, necessity and freedom in transformations of varieties is also discussed in [R4.10], [R4.13], [R1717].

4.1.3 Info-dynamics and a Prelude to the Development of a Unified Theory of Engineering Sciences: Socio-natural Technological Analytics

From the information space, a knowledge space is developed in the epistemological space to create the necessity and freedom in the social transformation space which contains social actual-potential polarities with residing dualities. This social transformation space includes all individual and collective decision-choice systems, social institutions, all forms of engineering, socio-physical command-control systems, all of which constitute a family of transformation decision-choice systems regarding the behavior of the family of actual-potential polarities. The elements from the knowledge space are the inputs into the social transformation decisions in relation to the family of the social actual-potential polarities. The construction of the knowledge space is continuous and never ending in the sense that the knowledge stock-flow process like the info-stock-stock flow process is always in disequilibrium dynamics. There are many sub-spaces in the social transformation space which is the space where cognitive agents can create and transform varieties by mimicking the laws of natural transformations in the ontological space.

The sub-spaces of social transformation correspond to the sub-spaces of knowledge, where the elements of sub-spaces of knowledge become inputs into social transformation decisions to destroy existing varieties and to create new varieties in accordance with either individual or collective preferences depending on the actual-potential relation of the preferences and specific needs requiring changes. An example of the social transformation space is the space of engineering of all forms such as electrical engineering, construction engineering, biomedical engineering, social engineering and others which are too many to name. The elements in the engineering space is unified and integrated by a unified theory of engineering sciences. The unified theory of engineering sciences requires a reasonable knowledge on existing or non-existing varieties which constitute the actual that may be destroyed and the potential varieties which may be transformed as a replacements and with a family of actual or potential specific technologies needed for of transformation.

4.1.3.1 Conditions of Technology of Socio-natural Transformations of Varieties

The problem of social transformations requires three types of knowledge. One type is a knowledge about the identity of the actual variety as well as the corresponding knowledge on its cost-benefit configuration. The second type of knowledge is about the identities of potential replacement varieties as well as the distribution of their cost-benefit configurations. The third type is the knowledge on technology of natural transformations which are engendered by both inter-categorial and intra-categorial conversions in inter-categorial and intra-categorial spaces in terms of Philosophical Consciencism with decision-choice intentionalities and implementation actions. To understand the transformation potential of an actual variety, the three knowledge systems must be combined. The knowledge of *what there is* (actual variety) and *what would be* (the potential replacement varieties) define the necessity and the corresponding necessary conditions for the destruction of the existing variety to maintain the info-stock and the replacement by new variety to create info-flow to update the info-stock. The knowledge about how to destroy the actual and how to construct the potential is the technology that define the freedom and corresponding sufficient conditions for transformation. The technology required for affecting and effecting variety transformations is made up of physical and social technologies, where the physical technology is completely and uncompromisingly dependent on the structure of the social technology. The information-knowledge conditions of the actual and potential varieties in relation to the family of natural polarities and transformation necessity have been discussed in this monograph and [R4.10], [R4.13], [R17.15].

It is useful, now, to turn an attention on the conditions of knowledge on technology and the corresponding social decision-choice systems that provide a solution to the problem of freedom to define the sufficient conditions for transformations of varieties to satisfy the general transversality conditions over the inter-categorial and intra-categorial spaces of the family of varieties and categorial varieties. For each actual variety to be destroyed, there are many potential varieties that may be actualized as a new variety. The transformation path to these potential varieties are many and defined by a family of sets of technologies. The choice of a variety from the potential space in the substitution-transformation process may be guided by a cost-benefit rationality which relates to necessity and freedom conditions. Similarly, for each potential variety that qualifies for actualization, there is a family of actual-potential technologies that can define the social freedom and the action to construct the sufficient conditions for the implementation and actualization in the dynamics of the family of the actual-potential polarities. The choice of an optimally appropriate technology may be guided by cost-benefit rationality given the socio-physical technological space.

Under the presence of necessity and necessary conditions as well as freedom and sufficient conditions one may understand that both inter-categorial and intra-categorial conversions are action varieties defined over the inter-categorial and intra-categorial spaces. What are the information conditions specified by characteristic-signal dispositions that define their differences and similarities. The characteristic-signal disposition suggests that the concept of variety transformation in inter-categorial space is about qualitative disposition properly defined. The concept of variety transformation in intra-categorial space is about quantitative disposition properly defined. The differences and similarities as established between the inter-categorial space and intra-categorial space must be properly related to inter-categorial conversion and intra-categorial conversion. The necessary conditions and necessities are different for inter-categorial and intra-categorial categorial conversions. The differences and similarities find expressions in the social transformation technological space that has been mastered from and related to the natural transformation technology of the dynamics of matter-energy-information process with neutrality of time. The inter-categorial space is the space where the internal transformation technologies act on the internal arrangement of the organizational structure of an element and transform its qualitative characteristic disposition, irrespective of the nature of its quantitative disposition, and hence its identity and move it into another category. Alternatively, the intra-categorial space is the space where the internal technologies act on the internal arrangement of the organizational structure of an element to maintain its qualitative disposition and identity within the same category but transform its quantitative characteristic disposition.

4.1.3.2 The Nature of Universal Technologies and Representations in Socio-natural Transformations of Varieties

In the information-knowledge process, two technological spaces are identified relative to the information production which is the destruction and creation of varieties. There is the natural technological space and the non-natural technological space. The information generation system is such that there is the natural (N = natural) technology space $\mathfrak{T}^{\mathrm{N}}$ with a generic element $t^{\mathrm{N}} \in \mathfrak{T}^{\mathrm{N}}$ which is composed of two natural sets of technologies of a set of natural qualitative technologies $\mathfrak{T}_{\mathrm{Q}}^{\mathrm{N}}$($\mathrm{Q}$ = qualitative) with a generic element $t_{\mathrm{Q}}^{\mathrm{N}} \in \mathfrak{T}_{\mathrm{Q}}^{\mathrm{N}}$ to act on an element and transform its qualitative disposition by destroying the existing qualitative characteristics and use them as inputs to create new qualitative characteristics while keeping it quantitative disposition unchanged. There is also a set of quantitative technologies $\mathfrak{T}_{\mathrm{R}}^{\mathrm{N}}$($\mathrm{R}$ = quantitative) with a generic element $t_{\mathrm{R}}^{\mathrm{N}} \in \mathfrak{T}_{\mathrm{R}}^{\mathrm{N}}$ to act on an element and transform its quantitative disposition by destroying the existing quantitative characteristics and use them as inputs to create new quantitative characteristics while keeping its qualitative disposition unchanged. The natural technological space is such that $\mathfrak{T}^{N} = \left(\mathfrak{T}_{Q}^{N} \bigcup \mathfrak{T}_{R}^{N}\right)$. It is possible that both qualitative and quantitative technologies may be simultaneously active on the same element to transform both its qualitative and quantitative dispositions and hence $\left(\mathfrak{T}_{Q}^{N} \bigcap \mathfrak{T}_{R}^{N}\right) \neq \varnothing$. The natural technologies are defined in the ontological space in terms of natural creative-destructive processes. Complementing the natural technological space in the universe of variety destruction-creation process is the non-natural technological space $\mathfrak{T}^{\not\aleph}$, ($\not\aleph$ = non-natural) the progress of which relates essentially to necessity and freedom of cognitive agents.

Definitions 4.1.3.1 (*Natural and Non-natural Technologies* $(\mathfrak{T}^{N}, \mathfrak{T}^{\not\aleph})$)
A technology $\mathfrak{T}^{N}$ is said to be natural if its variety-transformation process is through the ingenuity of natural decision-choice activities embodied in the internal forces of nature. It is said to be non-natural, $\mathfrak{T}^{\not\aleph}$ if its variety-transformation process is through the ingenuity of cognitive agents' decision-choice activities mimicking natural technologies through the understanding of natural variety-transformation decision-choice activities.

Definition 4.1.3.2 (*Qualitative and Quantitative Natural Technologies*)
Natural technology $\mathfrak{T}_{Q}^{N}$ is said to be qualitative if its actions and forces are directed to alter the qualitative characteristics of varieties. It is said to be quantitative $\mathfrak{T}_{R}^{N}$ if its actions and forces are directed in changing the quantitative characteristics of varieties.

The non-natural technological space is defined in the epistemological space. In social transformation of varieties cognitive agents must acquire the information on the natural technologies and mimic the processes to create the needed social technology for the transformation of a particular social variety or a set of social varieties. In other words, every set of conditions of technology known to cognitive agents exists in a natural form. The bringing into an active use in a social setting is the challenge of cognitively technological artistry of different social systems. It is also on this basis that it is maintained in the theory of knowledge that no social setup has a monopoly over the elements of the knowledge space. Knowledge is acquired by organization of search and research to connect to essences of natural existence. This natural existence and the knowledge of its activities are opened to all ontological agents with cognitive abilities [R3.15], [R3.49], [R3.47], [R3.53], [R3.77], [R3.78], [R4.10], [R4.9].

From the knowledge of conditions from the natural technological space, one may define a non-natural technological space, $\mathfrak{T}^{\not\aleph}$ with a generic element $t^{\not\aleph} \in \mathfrak{T}^{\not\aleph}$ which is a non-natural creation. The non-natural technological space is contained in the natural technological space such that $\left(\mathfrak{T}^{\not\aleph} \subset \mathfrak{T}^{N}\right)$ and that at any moment of time the social technology space is a very small subset of the natural technology space $\left(\#\mathfrak{T}^{\not N} \lll \#\mathfrak{T}^{N}\right)$ where the symbol $(\lll)$ means substantially less than. The non-natural technological space is made up of subspace of qualitative technology $\mathfrak{T}_{Q}^{\not N}$ with a generic element $t_{Q}^{\not N} \in \mathfrak{T}_{Q}^{\not N}$, subspace of quantitative technology $\mathfrak{T}_{R}^{\not N}$ also with a generic element $t_{R}^{\not N} \in \mathfrak{T}_{R}^{\not N}$ such that the general social technological space is a union of qualitative and quantitative technological subspaces $\mathfrak{T}^{\not N} = \left(\mathfrak{T}_{Q}^{\not N} \bigcup \mathfrak{T}_{R}^{\not N}\right)$ with non-empty intersection of the form $\left(\mathfrak{T}_{Q}^{\not N} \bigcap \mathfrak{T}_{R}^{\not N}\right) \neq \varnothing$. From the point of view of social information-knowledge process, the natural technological space constitutes a non-natural technological potential space in the form $\mathfrak{T}^{\not N\Pi} = \left(\mathfrak{T}_{Q}^{\not N\Pi} \bigcup \mathfrak{T}_{R}^{\not N\Pi}\right)$ where $\mathfrak{T}_{Q}^{\not N\Pi} \subset \mathfrak{T}_{Q}^{N}$ (Π = potential) and non-natural actual technological space $\mathfrak{T}^{\not N\Gamma} = \left(\mathfrak{T}_{Q}^{\not N\Gamma} \bigcup \mathfrak{T}_{R}^{\not N\Gamma}\right)$ where $\mathfrak{T}_{R}^{\not N\Gamma} \subset \mathfrak{T}_{R}^{N}$ and (Γ = actual).

Definition 4.1.3.3 (*Qualitative and Quantitative Non-natural Technologies*)
A Non-Natural technology $\mathfrak{T}^{\not N\Gamma}$ is said to be qualitative in nature if its actions and forces alter the qualitative characteristics of varieties. It is said to be quantitative in nature $\mathfrak{T}_{R}^{\not N}$ if its actions and forces are directed in changing the quantitative characteristics of varieties.

Definition 4.1.3.4 (*Actual and Potential Non-natural Technologies*)
A Non-Natural technology space $\mathfrak{T}^{\not N\Gamma}$ is actual and is said to be either qualitative actual $\mathfrak{T}_{Q}^{\not N\Gamma}$ or quantitatively actual $\mathfrak{T}_{R}^{\not N\Gamma}$ in nature if the actions and forces of its elements are known to cognitive agents for altering either the qualitative or quantitative characteristics of varieties. It is said potential $\mathfrak{T}^{\not N\Pi}$ if the actions and forces of both of its elements of qualitative in nature $\mathfrak{T}_{Q}^{\not N\Pi}$ and quantitative in nature $\mathfrak{T}_{R}^{\not N\Pi}$ for changing either the qualitative or the quantitative characteristics of varieties exist as unknown to cognitive agents in the space of the potential.

The socially actual technological space constitutes the known elements of technological know-how for social transformations. The socially actual technological space is dynamic and continually expanding in relation to information-knowledge dynamics. It is also useful to note that $\mathfrak{T}^{N} = \left(\mathfrak{T}_{Q}^{\not N\Pi} \bigcup \mathfrak{T}_{R}^{\not N\Pi}\right) \bigcup \left(\mathfrak{T}_{Q}^{\not N\Gamma} \bigcup \mathfrak{T}_{R}^{\not N\Gamma}\right) = \left(\mathfrak{T}^{\not N\Pi} \bigcup \mathfrak{T}^{\not N\Gamma}\right)$ which simply say that the potential and actual social technological spaces is equal to the natural technological space irrespective of the nature of information-knowledge dynamics. For the dynamics of the info-production process of the activities of cognitive agents over the epistemological space, the non-natural technology, in the potential and actual forms may be partitioned into *physical technological set* and *social technological set.* The social technological set is institutional arrangements and organizational forms that allow the creation and destruction of physical technology, institutional arrangements, and the mimicries of nature, product creation in terms for variety transformations in accordance with the needs and wants on the basis of cognition and decision-choice actions where variety transformations are guided by individual and collective preferences in individual and collective decision spaces.

In understanding the natural and social technological dynamics, it may be kept in mind that the information-knowledge stock-flow process is always in disequilibrium. The information- knowledge stock-flow disequilibrium process may produce new natural and social technologies that correspond to new information flows, the destruction of existing varieties and the creation of new varieties with corresponding identities under the transformation-substitution principles. Let us refer to the natural and social transformation processes as induced by *natural* $\mathfrak{T}$-process and *social* $\mathfrak{T}$-process. The natural $\mathfrak{T}$-process are intimately connected with inter-categorial and intra-categorial conversions to define the necessary conditions of social transformations of social varieties, while the social $\mathfrak{T}$-process define the sufficient conditions for social transformation of social varieties. The natural $\mathfrak{T}$-process are constraints on the social $\mathfrak{T}$-process where the maximization of the effectiveness of the social $\mathfrak{T}$-process is the maximization of freedom subject to the information-knowledge constraints from the natural $\mathfrak{T}$-process which form the necessity to change. It is over the space of social $\mathfrak{T}$-process that Philosophical Consciencism with all types of intentionalities are forms under principles of preference ordering. It is under the same space that misinformation and disinformation are developed as facts to justify intentionality in the space of conflicts, wars, atrocities, peace and many others [R17.15].

4.1.3.3 A Note on Parent-Successor Social Technologies in Transformations of Varieties

Over the epistemological space, the non-natural technologies may be placed in fuzzy categories where each technological category is a variety-transformation specific in the sense of its utility in accomplishing a set of particular variety destruction-creation activities. The non-natural technological space for variety transformation and info-production is a collection of a family f set of technologies where each set is specific to accomplish a task of either variety destruction or variety creation. The conditions of fuzzy technological categories point to the existence of overlapping technological categories where a technology may serve more than one transformation task. Every non-natural technology resides in the primary-derived duality in the sense that every non-natural or natural technology is both a parent and offspring in the activities of variety transformations and hence info-production. It is said to be a parent if other technologies may be reduced to its conditions of existence. It is said to be an offspring if it can be constructed from a previously existed technology. The conditions of parent-offspring technological process is such that the technological process, just like the info-dynamics is always in stock-flow disequilibria.

The nature of the existence of the parent-offspring technologies is revealed by the stock-flow disequilibrium process. The natural $\mathfrak{T}$-process always serves as the primary category as well as parent to the non-natural $\mathfrak{T}$-processes. The access to the structure of the natural $\mathfrak{T}$-processes by cognitive agents is through the principle of acquaintance, learning, understanding and mimicking through exactness and

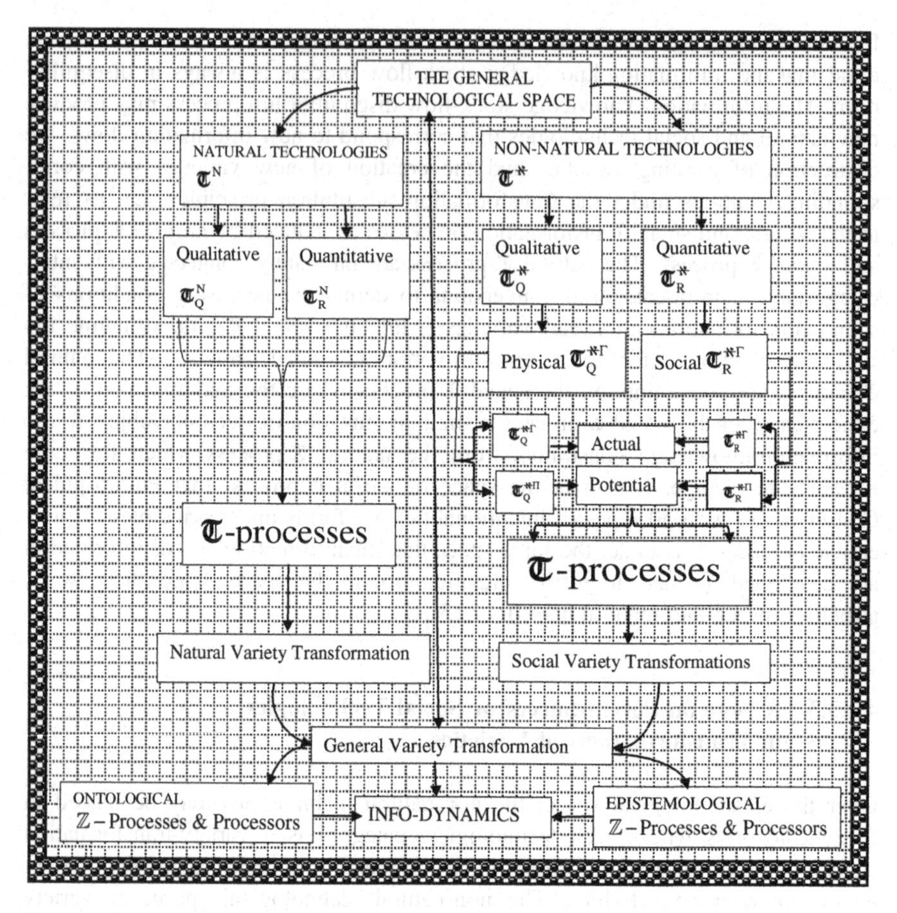

Fig. 4.1 An epistemic geometry of the paths of variety transformations and info-dynamics

inexactness processes The general process may be presented in a cognitive geometry as in Fig. 4.1.

4.2 The Theory of Info-dynamics and Categorial Conversions Under Conditions of Pure Time

The transformations of varieties over the ontological space with natural $\mathfrak{T}$-process and over the epistemological space with social $\mathfrak{T}$-process are through the dynamics of actual-potential polarities with matte-energy interplays. They are processes, the study of which requires an introduction of time-processes on the characteristic-signal dispositions that define the information at each time point with past-future continuity. The transformations are time-processors which act on the characteristic dispositions the changing identities of which are communicated through the signal dispositions. The introduction of time, therefore, is into the definition of information which requires that each of the building blocks must be time-indicated such that the object set, the set of phenomena, the characteristic disposition, the signal disposition and the set of varieties are equipped with a time set $\mathbb{T}$, where there are $(\Omega, \mathbb{T})$,$(\Phi, \mathbb{T})(\mathbb{X}, \mathbb{T}), (\mathbb{S}, \mathbb{T})$ as the object time set, the phenomenon-time set, the characteristics-time set and the signal-time set respectively, all of which correspond to the variety-time set, $(\mathbb{V}, \mathbb{T})$. The nature and the concept of the time set $\mathbb{T}$ have been discussed in Chap. 3 of this monograph where distinctions were made among *pure time*, *cost time*, *benefit time* and *cost-benefit time* in the transformation-decision process. The time set, whatever the concept and definition, must be related to a set of time-processors where each time processor corresponds to a time-point transformation of the same object in time-point varieties that allow the examination of operative behavior of categorial conversion and decision-choice intentionality defined in the *Consciencism* of the internal structure of the varieties [R17.16].

The structure of the transformation-decision process is the destruction of the existing varieties and the creation of new varieties in time-dependent variety space $(\mathbb{V}, \mathbb{T})$ and must be studied from the static and dynamic conditions as provided by $(\Omega, \mathbb{T})$, the time-dependent universal object set, $(\Phi, \mathbb{T})$, the time-dependent phenomenon space, $(\mathbb{X}, \mathbb{T})$,the time-dependent characteristic space, and $(\mathbb{S}, \mathbb{T})$ the time-dependent sign space. The static conditions involve in providing the definitional conditions of information on $(\mathbb{V}, \mathbb{T})$ which then establish the initial conditions of change, while the dynamic conditions provide transformation conditions on the primary universal set of varieties $(\mathbb{V}, \mathbb{T})$ with an info-support $(\mathbb{Z}, \mathbb{T})$ to a new universal set of derived varieties $(\mathbb{V}, \mathbb{T})_\pi$ with info-support $(\mathbb{Z}, \mathbb{T})_\pi$ where the subscript (π) identifies the derived.

The universal set of varieties $(\mathbb{V}, \mathbb{T})$ is said to be deformed at a future time if $\left[\#(\mathbb{V}, \mathbb{T})_\pi < \#(\mathbb{V}, \mathbb{T})\right]$ and it is said to be enhanced if $\left[\#(\mathbb{V}, \mathbb{T}) < \#(\mathbb{V}, \mathbb{T})_\pi\right]$. The information deformity implies conditions of variety destruction and creation but with a less than equal number of variety replacements in the transformation process as seen in terms of reduction in the number of varieties which may be viewed in terms of the process of extinction of some varieties. This involves some variety transformations without replacement. The information enhancement implies conditions of variety destruction, creation and mutation with more than equal number of variety replacements in the variety-transformation process. The concept of variety replacement or non-replacement is about quantitative disposition rather than qualitative disposition. The concept of variety transformation in the inter-categorial space is about the qualitative disposition, properly defined. The concept of variety transformation in the intra-categorial space is about the quantitative disposition, properly defined, where differences are established between the inter-categorial space and the intra-categorial space and properly related to the inter-categorial conversion and intra-categorial conversion.

The inter-categorial space is the space where the internal transformation technologies act on the internal arrangement of the organizational structure of an element and transform its qualitative characteristic disposition, irrespective of the nature of its quantitative disposition, and hence its identity and move it into another category. In other words, the original variety is destroyed from within with a technology of destruction and a new variety is created from within with a technology of creation leading to a creation of qualitative information. Alternatively, the intra-categorial space is the space where the internal technologies act on the internal arrangement of the organizational structure of an element to maintain its qualitative disposition and identity within the same category but transform its quantitative characteristic disposition without destroying its identity leading to the creation of a quantitative information. It is important to keep in mind that there is a general technology space $\mathfrak{T}$ with a generic element $t \in \mathfrak{T}$ which is composed of two sets of technologies of a set of qualitative technologies $\mathfrak{T}^{Q}$ with a generic element $t_{Q} \in \mathfrak{T}^{Q}$ to act on an element and transform its qualitative disposition, and a subset of quantitative technologies $\mathfrak{T}^{R}$ with a generic element $t_{R} \in \mathfrak{T}^{R}$ to act on an element and transform its quantitative disposition while keeping its qualitative disposition unchanged. The technological space is such that $\mathfrak{T} = \left(\mathfrak{T}^{Q} \bigcup \mathfrak{T}^{R}\right)$. It is possible that both qualitative and quantitative technologies may be simultaneously active on the same element to transform both its qualitative and quantitative dispositions and hence there is the condition where the intersection is nonempty $\left(\mathfrak{T}^{Q} \bigcap \mathfrak{T}^{R}\right) \neq \varnothing$. With respect to the introduction of the necessary conditions and sufficient **conditions of** socio-natural transformations of varieties, it may be helpful to explicitly incorporate the technological conditions into the object space $(\Omega, \mathbb{T}, \mathfrak{T})$, the space of phenomena $(\Phi, \mathbb{T}, \mathfrak{T})$ the space of varieties $(\mathbb{V}, \mathbb{T}, \mathfrak{T})$, the characteristic space $(\mathbb{X}, \mathbb{T}, \mathfrak{T})$, and the signal space $(\mathbb{S}, \mathbb{T}, \mathfrak{T})$. The nature of information and knowledge regarding the variety-transformation process must be specified regarding natural and social dynamics.

4.2.1 Transformation Operators and Introduction to the Theory of Info-dynamics

It must be analytically clear that socio-natural transformation is the production of information through the conversions of varieties and categorial varieties of systems. This is another way of seeing the concept of systems dynamics in the most complex form. The dynamics is about a continual production and storage of information. The complexity is about the size of the variety space that presents a multiplicity of problems of variety identification over an expansive fuzziness in information processing. The study of information is, thus, the study of systemicity in either automatic control or non-automatic control framework at both static and dynamic conditions. The theory of static systemicity is embedded in the theory of info-statics which has been presented in [R17.17]. The theory of dynamic systemicity is embedded in the theory of info-dynamics which is currently under development.

Central to the study and development of the theory of systems dynamics and complexity are time t with time set $(t \in \mathbb{T})$ and (time-transformation processor) *categorial-conversion operator* $\pi(t)$ with a set of time-transformation processors or categorial-conversion operators $\pi(t) \in (\Pi, \mathbb{T})$ that act on the elements of the key definitional building blocks of the information. The categorial-conversion operator is a function of the internal technologies that have been discussed above such that $\pi(t) \in (\Pi, \mathbb{T}) \subset \mathfrak{T}$. Every time set has special properties that allow any time-processor to belong to the set of time-processors. Few questions may be asked that will provide entry and exist points to the development of the theory of info-dynamics in support of the theory of info-statics. What is the structural relation between the set of conditions of info-statics and those of info-dynamics? Is the structure of the evolving transformation isomorphic to the structure of info-dynamics? Is the cost-benefit structure involved in this info-dynamic process? If cost-benefit is involved, what is its time-processor? If the costs and benefits are involved, then what are the relationships of costs to benefits in transformations? The results of the systems dynamics will be shown to generate an *organic information enveloping* $\mathbb{Z}_{\Omega}(t)$ of stock-flow conditions where $(t \in \mathbb{T})$. At any time-point $(t \in \mathbb{T})$, the organic information $\mathbb{Z}_{\Omega}$ is a static structure of the set of the form:

$$\mathbb{Z}_{\Omega} = (\mathbb{X} \otimes \mathbb{S}) = \{\mathbb{Z}_{v} = (\mathbb{X}_{v} \otimes \mathbb{S}_{v}) | v \in \mathbb{V}, \mathbb{X}_{v} \subset \mathbb{X}, \& \mathbb{S}_{v} \subset \mathbb{S}\} \tag{4.1}$$

The organic information enveloping is a set of individual variety-information enveloping $\mathbb{Z}_{v}$ of a particular variety $(v \in \mathbb{V})$ such that $\mathbb{Z}_{\Omega} = \bigcup_{v \in \mathbb{V}} \mathbb{Z}_{v}$ and $\mathbb{Z}_{\Omega} = \bigcup_{v \in \mathbb{V}} (\mathbb{X}_{v} \otimes \mathbb{S}_{v})$ which describes the total information space of all available varieties at any given time point. The space of varieties is the collection of all the individual varieties of the form $\mathbb{V} = \{v_i \,|\, i \in \mathbb{I}^{\infty} \text{ for any } t \in \mathbb{T}\}$. This is the conditions of info-statics, the inner structure of which must be examined. The info-statics presents a general information definition which is made up by characteristics that define the contents of the varieties. The characteristics of the varieties send signals that reveal their inner structures and hence their identities.

4.2.2 *The Structural Analytics of Characteristic-Signal Dispositions of Varieties in Complexity of Forces Under Conflicts*

To construct a theory of info-dynamics as a theory of the dynamic behavior on a system of transformation of varieties in its general form, it is useful to initialize the definitional conditions of the space of varieties that establishes the universal object set Ω, where every $\omega \in \Omega$ is a variety $v \in \mathbb{V}$. Each variety has a corresponding characteristic-signal disposition $\mathbb{Z}_{v} = (\mathbb{X}_{v} \otimes \mathbb{S}_{v})$ and corresponding phenomenon

$\phi_v \in \Phi$. The definitional conditions of information at the static state must be the same and must hold for each element in the space. The inner structure of each variety will vary in accord with the information content $\mathbb{X}_v$ and corresponding signals $\mathbb{S}_v$. Every categorial variety is a collection of the same varieties which is defined by a characteristic set $\mathbb{X}_{\mathbb{C}}$ referred to as a categorial characteristic disposition, where $\mathbb{C}_v$ is the home category of $v \in \mathbb{V}$. Corresponding to each categorial characteristic disposition is a categorial signal disposition $\mathbb{S}_{\mathbb{C}}$. The two together define information on the category of the same varieties $\mathbb{V}_{\mathbb{C}}$. The internal conditions of these categorial varieties are under plenum of forces from opposing forces that are generated under the principle of opposites composed of infinite set of socio-natural dualities which supports an infinite set of socio-natural polarities to produce energy from within and to generate internal self-motions of varieties and categorial varieties. Here, the universe is considered as a unit composed of internal opposing parts for conflicts with relation destruction and creation. The theory of info-dynamics is a general theory of dynamics of elements in the universal object set in internal unity and relational continuum.

The characteristic disposition is the same for all varieties in a fixed category and is composed of two opposite sub-sets of negative characteristics set $\mathbb{X}_{\mathbb{C}}^{N}$ with a corresponding negative signal set $\mathbb{S}_{\mathbb{C}}^{N}$, and positive characteristics set $\mathbb{X}_{\mathbb{C}}^{P}$ with a corresponding positive signal set $\mathbb{S}_{\mathbb{C}}^{P}$. The set conditions are such that the information about each categorial variety of the characteristic-signal disposition $\mathbb{Z}_{\mathbb{C}} = (\mathbb{X}_{\mathbb{C}} \otimes \mathbb{S}_{\mathbb{C}}), \mathbb{X}_{\mathbb{C}} = (\mathbb{X}_{\mathbb{C}}^{P} \bigcup \mathbb{X}_{\mathbb{C}}^{N})$ and $(\mathbb{X}_{\mathbb{C}}^{P} \bigcap \mathbb{X}_{\mathbb{C}}^{N}) \neq \varnothing$ with $\left(\#\mathbb{X}_{\mathbb{C}}^{P} \lesseqqgtr \#\mathbb{X}_{\mathbb{C}}^{N}\right)$ are supported by $\mathbb{S}_{\mathbb{C}} = (\mathbb{S}_{\mathbb{C}}^{P} \bigcup \mathbb{S}_{\mathbb{C}}^{N})$ and $(\mathbb{S}_{\mathbb{C}}^{P} \bigcap \mathbb{S}_{\mathbb{C}}^{N}) \neq \varnothing$, with $\left(\#\mathbb{S}_{\mathbb{C}}^{P} \lesseqqgtr \#\mathbb{S}_{\mathbb{C}}^{N}\right)$, in order to establish its identity and the category to which a variety belongs. Similarly, $(\mathbb{X}_{\mathbb{C}} \otimes \mathbb{S}_{\mathbb{C}}) == \left((\mathbb{X}_{\mathbb{C}}^{P} \otimes \mathbb{S}_{\mathbb{C}}^{P}) \bigcup (\mathbb{X}_{\mathbb{C}}^{N} \otimes \mathbb{S}_{\mathbb{C}}^{N})\right)$ and $\mathbb{Z}_{\mathbb{C}} = (\mathbb{Z}_{\mathbb{C}}^{P} \bigcup \mathbb{Z}_{\mathbb{C}}^{N})$. The set specifications establish the necessary conditions that create *external relations* for transformation of varieties and corresponding identities. These opposites establish conditions of active dualities with positive dual and negative dual that generate internal decision-choice activities opposing to each sub-set to create conflicts and tensions for every categorial variety. The conflict activities of the duals of the duality create a continual production of energy for forces that produce work of internal instabilities with a temporary qualitative equilibrium and generate the necessary conditions for internal change. Every element in the universal object set is in information stock-flow disequilibrium. The temporary qualitative equilibrium states of elements apply to the transformation decision time which is basically a period of pure time. The necessary conditions relate to external environment in relational continuum; the sufficient conditions relate to arrangements of internal characteristics with intentionality to self-transformation.

It is here that the claim of every variety is a set of forces under tension finds an epistemic justification. It is also here that the statement that sustainable transformations are only possible from within and not from without finds analytical support. The continual production of energy is under opposing intentionalities of the relative negative positive characteristic sub-sets where the positive characteristic sub-set seeks a dominance over the negative characteristic subset either to retain or change the existing variety and establish a new identity with a new information in accord with the positive dual, and the negative characteristic sub-set seeks a dominance over the positive sub-set either to retain or change the existing variety and establish a new identity with a new information in accord with the intentionality of the negative dual [R17.15]. It easily follows that Sub-categories may be formed within the qualitative categories that define inter-categorial varieties on the basis of quantitative disposition to establish intra-categorial varieties. The important thing about the structure is that a sustainable transformation comes from within the internal constitutions of the varieties. The sustainability of social progress in the fields of economics, politics and law finds find its logical support from the concept of internal transformation through internal forces of varieties.

Both the ontological space and the epistemological space are infinite in existence and are partitioned by the categories of being. These categories of being are expressions of existential varieties in similarity and difference that allow comparative analytics in the epistemological space by cognitive agents. The epistemological categories are inexactly known to cognitive agents through the acquaintances of the signal dispositions while a small set of the ontological categories are only accessible to cognitive agents due to all forms of cognitive limitationality and limitativeness in the source-destination duality of ontological-epistemological relational structure. Over the epistemological space, the category formation is essential in the development of a language with the corresponding vocabulary and grammar of thought through the principle of *nominalism* [R17.17]. This is made possible by acquaintances where the acquaintances may be enhanced by technical knowhow. The size of vocabulary expands as new ontological varieties and categorial varieties are known and new epistemological varieties are created by cognitive agents. In simple terms, the ontological space is filled with objects, things, states and processes which may be grouped into categories of difference and commonness by means of their qualitative characteristics. Language in any form is simply a representation, and communication of information about varieties. The characteristics of varieties and categorial varieties present themselves as information through the signals to cognitive agents operating in the epistemological space.

Given the nature of the representation, the cognitive agents must process this information with *epistemic tools* to try to find an isomorphism between the ontological elements in the ontological space and epistemological elements in the epistemological space to claim knowledge as the contents from the information to be used in the activities of communication, teaching and learning. The epistemic tools are couched in methodological constructionism and reductionism relative to a paradigm of thought. The foundational discussions on these critical concepts and their roles in reason and thought may be found in [R3], [R3.7], [R3.10], [R3.13], [R3.18]. Every name and every concept in the vocabulary must meet the qualitative characteristic conditions of a category where every inter-category is distinguished from others by its qualitative disposition and every intra-category is distinguished from within by quantitative disposition given the qualitative disposition. Alternatively viewed, a vocabulary is a set of qualitative-quantitative dispositions in nominalism that creates foundation for the application of methodological constructionism and reductionism.

This methodological duality with relational continuum and unity is the same for all knowledge constructions the input of which is an abstraction from the signal disposition through acquaintances. The methodological constructionism provides us with an epistemic toolbox for the development of explanatory and prescriptive science and theories to derive knowledge over the epistemological space. The methodological reductionism provides us with an epistemic toolbox to verify claims of knowledge by tracing it to the primary category which is the characteristic disposition. The methodological reductionism encompasses the principles of verification, falsification, justification, corroboration and others [R4.10], [R3.11], [R3.39], [R3.52], [R3.53], [R3.64], [R3.79], [R7.21], [R8.41], [R3.67], [R3.68]. The concept of category defined in terms of common characteristic set is based on the implicit structure of *characteristic-theory of categorial analytics* which requires the characteristics of each object to be identified and then placed in an appropriate category for identity distinction that locates common varieties. This is the *content identification problem of information*. This information problem is the *general identification problem* in sciences, non-sciences, knowing and the development of knowledge systems. It is also based on *signal-theory of categorial analytics* which requires that every characteristic of objects must have a corresponding signal in communication for the characteristic identification of varieties. This is the *communication problem* of information contents. In this process, there are *characteristic theory* and *signal theory* that must be complemented by a *theory of transformation technology* in understanding the theories of info-dynamics and info-statics, where every transformation of a variety is a production of information flow and a maintenance of individual and collective information stocks. The general identification problem, the problem of preference ordering, the problem of optimal decision-choice actions, the problem of language development, the problem of distinction and similarities and many others are all defined over the epistemological space and find meaning and utility under the principles of varieties and categorial varieties with defined identities supported by the distribution of characteristic-signal dispositions.

Supporting the characteristic-theory of categorial analytics is the signal-theory of categorial analytics under the principle of acquaintance. Over the epistemological space and under the principle of acquaintance, the signal theory provides the inputs for derived outputs of informing and knowing through the methodological constructionism for any given paradigm of thought. The results of informing and knowing become inputs into the mechanism of *characteristic analytics* through methodological reductionism under a given paradigm of thought. The characteristics form the primary category while the signals form the derived category. The primary and derived categories are about the contents of information that distinguish varieties in the process of informing and knowing. The validity of any derived category of informing and knowing is checked by the use of the set of methods and techniques of methodological reductionism. In other words, the methodological constructionism and reductionism constitute a methodological duality in a logically relational continuum and unity, where every theoretical element of constructionism has a supporting theoretical element of reductionism just like every element of truth has a corresponding element of falsehood support and vice versa and every element of good has a corresponding element of bad support in the judgment space as viewed from the fuzzy logical table and the supporting paradigm of thought [R4.7], [R4.8]. The material epistemic core of the constructionism-reductionism structure may be seen in the parent-offspring transformation relations, the past-present-future continuity, the self-creation-destruction structural dynamics, the self-correction and self-learning processes within the general information sock-flow disequilibrium dynamics of the primary and derived categories of matter and energy in the universal system of relational continuum and unity. It is the understanding of this material epistemic core that provides a powerful learning framework to mimic nature's creative-destruction processes to empower the creative essence of cognitive agents in the fields of organization of social systems in terms of production of science and engineering in the creation and destruction of varieties over the epistemological space.

Over the ontological-epistemological spaces there is an important conceptual idea about the nature of information that has been projected here in the theory of info-dynamics and the works that have been provided in the theory of info-statics [R17.17]. At the level of info-statics information is a set of relations among things at each pure time-point, cost-time point and benefit-time point. Information as a set of relation among things establishes the meaning of life as a set of solution-problem conditions in relations to things within the set of cost-benefit dualities. The set of relations at any pure time point allows the establishment of relational continuum and unity in the universal order under temporary information stock-flow equilibria which also establishes temporary solution-problem equilibria of relationality among things. At the level of info-dynamics, the information flow is the set of changing relations among things at infinite time in pure, cost and benefit conditions. The set of changing relations establishes a set of changing solution-problem conditions among things in the universal order in infinite time relative to pure time, cost time and benefit time. In other words, the transformations of varieties are transformation of information relations among things to give a continual relational dynamics as well as transformation of solution-problem conditions among things.

4.2.3 *The Foundational Analytics of the Cognitive Blocks for the Theory of Socio-natural Transformation and Information Production*

The epistemic foundations to the understanding of the cognitive structure being developed in this monograph to present a theory of info-dynamics relate to some important key concepts for the logical construct. These key concepts which will be defined and explicated in the later discussions are category, primary category, derived category, polarity, actual-potential polarity, duality, qualitative-quantitative duality, negative-positive duality, negative-positive characteristic sets, internal conflicts, internal force-energy structure, socio-natural decision-choice system, matter-energy self-excitement, paradigms of thought and principle of opposite. The general principle of opposites encompasses all these conceptual building blocks on the basis of which varieties find universal existence. These key concepts are in relation to static and dynamic conditions of internal constitutions of matter-energy structures. Matter is taken as the *primary category* which presents itself in many different forms called varieties of matter. Relationally, every element of matter has identity and constitutes a primary category of a variety which is related to states, processes, events, outcomes and phenomena in varieties. Similarly, energy is a derived category which presents itself in many different forms from matter. Relationally, every element of energy has identity and this identity constitutes a derived category of a variety which is related to states, processes, events, outcomes, communication and phenomena in varieties. Thus, one can speak of varieties in different forms which constitute a family of varieties or a family of categories. The formation of categories by indicator function has been discussed in Chap. 3 and with an extensive framework in [R17.15]. The categorial indicator function will be related to fuzzy indicator function in the possibility and the probability spaces in an attempt to understand necessity-freedom relationality under transformations and decision-choice uncertainties.

The concept of varieties may be taken as a concept of *species*. The identity of every variety at any time point is revealed by the corresponding characteristic-signal disposition. The collection of all characteristic-signal dispositions of the same variety from the past to any time point provides a total information available on the variety at that time point which will be the present information stock. The collection of all characteristic signal-dispositions of different varieties or categorial varieties from the past to any time point provides the total information available about the object space, the space of phenomena and the space of varieties at that time point which will be the total present information stock in the total system. The collection of all the varieties, the initial time of which may be arbitrarily specified relative to a to present, is the set of the initial static conditions for the future info-dynamic process, as seen in terms of the total matter-energy-information structure. All other categories and categorial varieties such as energy and forms of energy are internally matter-derived. Initial matter and the collection of initial varieties present conditions to study socio-natural transformations. Information is a set of relations among the primary and derived varieties, the collection of which constitutes socio-natural system in complexity and dynamics. These information relations define the identities of the varieties in terms of difference and commonness in continual transformations that produce time-point derived varieties. The theory of info-dynamics, therefore, is about the study of destruction and creation of varieties and the continual creation of information flow and information accumulation.

Every derived variety is seen along the time trinity of the past-present-future structure of transformations. In other words, every variety is time-dated and hence its information is also time-dated. Every variety at every time point is of the form

$v_\ell(t) \in \mathbb{V}(\mathbb{T})$ with a corresponding information structure in the form $\mathbb{Z}_\ell(t) = (\mathbb{X}_\ell(\mathbb{T})) \otimes (\mathbb{S}_\ell(\mathbb{T}))$ where $t \in \mathbb{T}$ and $\ell \in \mathbb{I}^\infty$. The term $\mathbb{Z}_\ell = (\mathbb{X}_\ell \otimes \mathbb{S}_\ell)$ is a stock of information on $v_\ell \in \mathbb{V}$ at any given time $t \in \mathbb{T}$, and the term $\mathbb{Z}_\ell(t) = (\mathbb{X}_\ell(\mathbb{T})) \otimes (\mathbb{S}_\ell(\mathbb{T}))$ is an info-production which is the stock-flow information process for each variety $v_\ell \in \mathbb{V}$. The stock-flow information processes are transformations of varieties as well as the changing of relations among varieties, where the dated same element $\omega_\ell(t) \in \Omega(\mathbb{T})$ and phenomenon $\phi_\ell(t) \in \Phi(\mathbb{T})$ present a time-dated variety, $v_\ell(t_1) \in \mathbb{V}(\mathbb{T})$, the enveloping path of which may be studied in terms of time differences and similarity. In other words $v_\ell(t) \in \mathbb{V}(\mathbb{T})$ is compared to $v_\ell(t_2) \in \mathbb{V}(\mathbb{T})$ in order to conclude whether if $t_2 - t_1 > 0$ then $v_\ell(t_1) \approx v_\ell(t_2)$ is identical and hence there is no transformation of $v_\ell \in \mathbb{V}$ between $t_1, t_2 \in \mathbb{T}$ and hence the information support $\mathbb{Z}_\ell(t_1) = \mathbb{Z}_\ell(t_2) = (\mathbb{X}_\ell(\mathbb{T})) \otimes (\mathbb{S}_\ell(\mathbb{T}))$ where $(\approx)$ is an identicality relation.

In the case of non-transformation of a variety $v_\ell \in \mathbb{V}$, the information conditions are such that $\mathbb{Z}_\ell(t_1) \subseteq \mathbb{Z}_\ell(t_2) \subseteq \mathbb{Z}_\ell(t_1)$. Even though the varieties at time $t_1, t_2 \in \mathbb{T}$ have not changed there is an information flow between these time points and hence $\left(\bigcup_{t=-\infty}^{t_1} \mathbb{Z}_\ell(t_1)\right) \subset \left(\bigcup_{t=-\infty}^{t_2} \mathbb{Z}_\ell(t_2)\right)$ which simply means that the stock of information at time $t_1 \in \mathbb{T}$ is contained in the stock of information at time $t_2 \in \mathbb{T}$ where $t_2 - t_1 > 0$. Alternatively, if the time-dependent varieties at $t_1, t_2 \in \mathbb{T}$ are non-identical $v_\ell(t_1) \not\approx v_\ell(t_2)$ in which case there is a transformation of $v_\ell(t_1) \rightarrow v_\ell(t_2)$ with a differential information support, where such an information support is in the form $\mathbb{Z}_\ell(t_1) = (\mathbb{X}_\ell(\mathbb{T})) \otimes (\mathbb{S}_\ell(\mathbb{T})) \neq \mathbb{Z}_\ell(t_2) = (\mathbb{X}_\ell(\mathbb{T})) \otimes (\mathbb{S}_\ell(\mathbb{T}))$ for the variety $v_\ell \in \mathbb{V}$ and with the condition that there is a differential information flow, where $\mathbb{Z}_\ell(t_1) \neq \mathbb{Z}_\ell(t_2)$, however, $\left(\bigcup_{t=-\infty}^{t_1} \mathbb{Z}_\ell(t_1)\right) \subset \left(\bigcup_{t=-\infty}^{t_2} \mathbb{Z}_\ell(t_2)\right)$. This simply means that the stock of information at time $t_1 \in \mathbb{T}$ is contained in the stock of information at time $t_2 \in \mathbb{T}$, where $t_2 - t_1 > 0$. In both cases, the stocks of information have increased but distinguished by the flow conditions.

The theory of info-dynamics is the explanatory and prescriptive processes of the collection of all the results of the socio-natural transformations. It is also about the enveloping of dynamics of any given variety with a corresponding information induced by either internal or external categorial conversions in a manner where the identity of a variety at any given time point is revealed by the corresponding information set. It is important to keep in mind that matter and energy present themselves in varieties and categorial varieties. These varieties and categorial varieties have identities that can only be revealed by the underlying information structures in varieties, where the contents are the characteristic dispositions and the awareness, transmission and communication are done through the signal dispositions. These information structures contain essential elements of *contents* that send different *signal sets* for distinction and similarity to create conditions of knowability of the inner essence of the identities of the varieties.

Here, it may be noted that the multiplicity of areas of cognitive activities and epistemic studies is due to the multiplicity of matter-energy varieties in static and dynamic information relations which separate and unite the universal objects in continuum. It is this continuity process of the universe that led FQXi: Foundational

Questions Institute to pose the question: "Is reality digital or analog?" Similarly, it is the epistemic conflict between existence and representation that also led FQXi to pose the question: "It from Bit or Bit from It?" [www.FQXi.org]. In both questions, the types of answer relates to the mind-body problem in the sense of where one place oneself in the primary-derived conceptual knowledge-ignorance polarity. My answers to these two important fundamental questions can be found in the same site [www.FQXi.org].

4.2.4 The Information Space and Info-algebras in Socio-natural Transformations

The information space is the collection of all the characteristic-signal dispositions of varieties in static and dynamic conditions in an infinite order, where socio-natural transformations of varieties take place. The static conditions help in solving the identification problem of varieties in time points while the dynamic conditions help in solving the changing identification problems of new varieties generated by system of socio-natural transformations. The process of socio-natural transformation of elements in the universal system is also the process of destruction and creation of varieties. The destruction and creation of varieties constitute a continual production of information, which allows same varieties to be defined and identified from within the time-structure of changing characteristic-signal dispositions which present the structure of the information space. The content space of the varieties is an infinite collection of the characteristic dispositions while the awareness space is an infinite collection of the signal dispositions. The content space and the awareness space constitute the information space. The dynamics of destruction and creation of information is here called the *info-dynamic processes* which are socio-natural transformations of varieties. The socio-natural transformations may be internally induced or externally imposed in specific forms of relational continuum and unity under the general principle of opposites with the game strategies of dominance. The relational differences of internally induced and externally imposed transformations are found in the sustainability process in relation to some concept of time which has been discussed in Chap. 3.

There is another important analytical core in the epistemic framework of the theory of info-dynamics in support of changing information relations among things and dynamics of the solution-problem structures of transformations. The important analytical core is that the categorial conversion part of general transformation places the information stock-flow disequilibrium in a game space for the continual destruction and creation of varieties and categorial, while the Philosophical Consciencism part of the general transformation places the information stock-flow disequilibrium process in the space of decision-choice strategies and tactics for continual destruction and creation of varieties and categorial varieties. The introduction of time allows one to define time-transformation operators as processors in terms of relational continuum and unity such that the time-transformation operators may be considered as $\mathbb{Z}$*-processors* (information processors). The relational continuum is established by some form of *information set connectivity* through containments such that the information stock on any variety $v \in \mathbb{V}$ at any two time points may be expressed as $\left(\bigcup_{t=-\infty}^{t_i} \mathbb{Z}_\ell(t_i)\right) \subset \left(\bigcup_{t=-\infty}^{t_j} \mathbb{Z}_\ell(t_j)\right) \forall (t_j - t_i) > 0 \in \mathbb{T}$, and $i, j \in \mathbb{N}$. The relational unity is established through some form of information stock overlapping in relational intersection for any variety $v_\ell \in \mathbb{V}$ at any two time points and may be expressed as $\left(\bigcup_{t=-\infty}^{t_i} \mathbb{Z}_\ell(t_i)\right) \bigcap \left(\bigcup_{t=-\infty}^{t_j} \mathbb{Z}_\ell(t_j)\right) \neq \varnothing \;\forall (t_j - t_i) > 0 \in \mathbb{T}$, and $i, j \in \mathbb{N}^*$ and hence $\mathbb{Z}(t_j) = \bigcup_{(t_j - t_i) > 0 \in \mathbb{T}} \mathbb{Z}(t_i)$. The information connectivity and unity require an appropriate relational connectivity and unity of time set $\mathbb{T}$ and variety space $\mathbb{V}$ in order to establish a process of transformation.

It is always useful to keep in mind that for a properly defined pure time set $\mathbb{T}$ without the conditions of cost, benefit and cost-benefit implications in transformation decisions, the enveloping path of each variety through the transformation process provides its information stock $\mathbb{Z}_v^*(t)$ and information flow $\mathbb{Z}_\ell(t)$ defining the variety's historic path of reality of the form $\mathbb{Z}_v^*(t) = \bigcup_{t=-\infty}^{t \in \mathbb{T}} \mathbb{Z}_\ell(t)$. These are information conditions that provide an input to knowledge production on each variety to affirm the idea that the study of a variety is the study of information. Given the enveloping path of each variety in the transformation of varieties, the total information stock of the universal object set $\mathbb{Z}_\Omega^*(t)$ at any given time point is $\mathbb{Z}_\Omega^*(t) = \bigcup_{v \in \mathbb{V}} \mathbb{Z}_v^*(t) = \bigcup_{v \in \mathbb{V}} \bigcup_{t=-\infty}^{t \in \mathbb{T}} \mathbb{Z}_\ell(t)$ and that $(\mathbb{X}(\mathbb{T})) \otimes (\mathbb{S}(\mathbb{T})) = \bigcup_{\ell \in \mathbb{I}^\infty(t)} (\mathbb{X}_\ell(\mathbb{T})) \otimes (\mathbb{S}_\ell(\mathbb{T}))$ which then corresponds to the object set $\Omega(\mathbb{T})$ with the phenomenon set $\Phi(\mathbb{T})$ that presents time-dated variety set $\mathbb{V}(\mathbb{T})$ as a unified system of enveloping realities at appropriately defined time set $\mathbb{T}$ where $\ell \in \mathbb{I}^\infty$ is and indexing of the varieties. It is analytically useful to observe that the total information stock is $\mathbb{Z}_\Omega^*(t) = \bigcup_{v \in \mathbb{V}} \bigcup_{t=-\infty}^{t \in \mathbb{T}} \mathbb{Z}_v(t)$ and the total information flow $\mathbb{Z}_\Omega(t)$ in the system at any $t \in \mathbb{T}$ is $\mathbb{Z}_\Omega(t) = \bigcup_{v \in \mathbb{V}} \mathbb{Z}_v(t)$. These are the information approaches to the study of statics and dynamics of systems in complexity.

Proposition 4.2.4.1 (Stock-Flow Condition)

The total information stock contained in the object space Ω at any time point is simply the union of all information stocks on all varieties that may be written as $\mathbb{Z}_\Omega^*(t) = \bigcup_{v \in \mathbb{V}} \bigcup_{t=-\infty}^{t \in \mathbb{T}} \mathbb{Z}_v^*(t)$ where $\mathbb{Z}_v^*(t) = \left(\bigcup_{t=-\infty}^{t \in \mathbb{T}} \mathbb{Z}_v(t)\right)$ is the accumulated information at time $t \in \mathbb{T}$ from the initial conditions of the primary variety or categorial variety at time $t = -\infty$ and $\mathbb{Z}_v(t)$ is the information flow of the variety $v \in \mathbb{V}$.

Proposition 4.2.4.2 (Monotonic Condition)

At any given time points $t_i, t_j \in \mathbb{T}$ such that $(t_j - t_i) > 0 \in \mathbb{T}$ and for any variety $v \in \mathbb{V}$, it is the case that $\left(\bigcup_{t=-\infty}^{t_1} \mathbb{Z}_v(t_1)\right) \subset \left(\bigcup_{t=-\infty}^{t_2} \mathbb{Z}_v(t_2)\right)$ and

$\left(\mathbb{Z}_{\Omega}(t_1) = \bigcup_{v\in\mathbb{V}} \bigcup_{t=-\infty}^{t_1\in\mathbb{T}} \mathbb{Z}_v(t_1)\right) \subset \left(\mathbb{Z}_{\Omega}(t_2) = \bigcup_{v\in\mathbb{V}} \bigcup_{t=-\infty}^{t_2\in\mathbb{T}} \mathbb{Z}_v(t_2)\right)$ *with* $\mathbb{Z}_v^*(t_1) \subset \mathbb{Z}_v^*(t_2)$ *and* $\mathbb{Z}_{\Omega}^*(t_1) \subset \mathbb{Z}_{\Omega}^*(t_2)$.

The complexity of the system is such that $\bigcup_{t=-\infty}^{t_i\in\mathbb{T}} \mathbb{Z}_v(t_i)$ is a horizontal union over time-dated information flow sets of any given variety while the second union is variety indexed sets of information over all column vectors $\bigcup_{v\in\mathbb{V}}(\cdot)$. Under these conditions it is useful to define an information space that accounts for stock-flow conditions where $\mathbb{Z}^*(t)$ represents info-stock and $\mathbb{Z}(t)$ represents info-flow.

Definition 4.2.4.1 (*The Information Space*)

The information space $\mathbb{Z}_{\Omega}^*(t)$ of the universal object space Ω, phenomenon space Φ and variety space $\mathbb{V}$ is the collection of individual variety information stocks $\mathbb{Z}_v^*(t)$ and information flows $\mathbb{Z}_v(t)$ for all varieties $v \in \mathbb{V}$ in the form $\mathbb{Z}_{\Omega}^*(t) = \{\mathbb{Z}_v^*(t) = \mathbb{X}_v \otimes \mathbb{S}_v | v \in \mathbb{V}, \forall t \in \mathbb{T}\}$ with $\mathbb{Z}_v^*(t) = \bigcup_{t=-\infty}^{t\in\mathbb{T}} \mathbb{Z}_v(t)$ and where there is a sequence of the form $\mathbb{Z}_v^*(t_i) \subset \mathbb{Z}_v^*(t_j)$, $\forall t_j > t_i \in \mathbb{T}$ and similarly, $\mathbb{Z}_{\Omega}^*(t_i) \subset \mathbb{Z}_{\Omega}^*(t_j), \forall t_j > t_i \in \mathbb{T}$ are time-point increasing sequences of information stock on individual variety and all varieties respectively.

The information space is extremely complex that defines and establishes relations of things. This is the meaning of complexity of systems since any system, no matter how it is defined is simply a set of relations of things. The information space is made up of two important variables of stocks and flows such that the individual stocks and the total stock are continually increasing in time dependent sequence where the flows of the individual varieties and the total flow are generated by destruction-creation processes of varieties in continual transformation where the information on the primary categorial varieties from $(t = -\infty)$ establishes the beginning that may be seen in terms of info-algebra and limits of information sets. The information space that has been constructed here is the most general and applicable to all areas of existing and potential knowledge systems. The info-stock conditions allow the study of static relations of things relative to the general solution-problem space. The info-flow conditions allow one to study dynamic conditions of things relative to the evolving general solution-problem space.

Definition: 4.2.4.2 (*Info-algebra*)

Given an information space $\mathbb{Z}_{\Omega}^*(t)$ the collection $\mathscr{B}(t)$ of all information stock-flow sub-sets of $\mathbb{Z}_{\Omega}^*(t)$ is an info-algebra if:

(i) $\mathscr{B}(t)$ is an infinite set at $t \in \mathbb{T} = (-\infty, +\infty)$ (expansiveness).
(ii) $\mathbb{Z}_{\Omega}^*(t), \mathbb{Z}_v^*(t) \in \mathscr{B}(t), \forall v \in \mathbb{V}$ (stock inclusion).
(iii) $\mathbb{Z}_{\Omega}(t), \mathbb{Z}_v(t) \in \mathscr{B}(t), \forall v \in \mathbb{V}$ (flow inclusion).
(iv) $\mathbb{Z}_{\Omega}^*(t_i) \subset \mathbb{Z}_{\Omega}^*(t_j) \in \mathscr{B}(t), \forall t_j > t_i \in \mathbb{T}$, (increasing monotonic total info-stock).
(v) $\mathbb{Z}_v^*(t_i) \subset \mathbb{Z}_v^*(t_j) \in \mathscr{B}(t), \forall t_j > t_i \in \mathbb{T}$, (increasing monotonic variety info-stock).

(vi) For a time-dated sequence of total stock $\mathbb{Z}^*_\Omega(t_i), \mathbb{Z}^*_\Omega(t_j) \in \mathscr{B}(t), \forall t_j, t_i \in \mathbb{T}$, it is the case that $\bigcup_{t=-\infty}^{t_j} \mathbb{Z}^*_\Omega(t_i) \in \mathscr{B}(t), \forall t_j > t_i \in \mathbb{T}$ (conditions of information accumulation).

(vii) For time-dated sequence of total stock $\mathbb{Z}^*_\Omega(t_i), \mathbb{Z}^*_\Omega(t_j) \in \mathscr{B}(t), \forall t_j, t_i \in \mathbb{T}$, it is the case that $\bigcap_{t=t_j}^{-\infty} \mathbb{Z}^*_\Omega(t_i) \in \mathscr{B}(t), \forall t_j > t_i \in \mathbb{T}$ (conditions of total primary information).

(viii) For a time-dated sequence of variety stock $\mathbb{Z}^*_v(t_i), \mathbb{Z}^*_v(t_j) \in \mathscr{B}(t), \forall t_j, t_i \in \mathbb{T}$, it is the case that $\bigcup_{t=-\infty}^{t_j} \mathbb{Z}^*_v(t_i) \in \mathscr{B}(t), \forall t_j > t_i \in \mathbb{T}$ (conditions of variety information accumulation and non-destructibility).

(ix) For a time-dated sequence of variety information of total stock $\mathbb{Z}^*_v(t_i), \mathbb{Z}^*_v(t_j) \in \mathscr{B}(t), \forall t_j, t_i \in \mathbb{T}$, it is the case that $\bigcap_{t=t_j}^{-\infty} \mathbb{Z}^*_v(t_j) \in \mathscr{B}(t), \forall t_j > t_i \in \mathbb{T},\ i, j \in \mathbb{N}^+$ (conditions of primary information on variety).

(x)
$$\left\{\begin{array}{ll} \mathbb{Z}^*_v(t_{-\infty}) = \bigcap_{t=t_j}^{-\infty} \mathbb{Z}^*_v(t_j) \text{ and} & \\ & \text{Primary and Derived conditions on variety dynamics} \\ \mathbb{Z}^*_v(t_j) = \bigcup_{t=-\infty}^{t_j} \mathbb{Z}^*_v(t_i) & \\ \mathbb{Z}^*_\Omega(t_{-\infty}) = \bigcap_{t=t_j}^{-\infty} \mathbb{Z}^*_\Omega(t_j) \text{and} & \\ & \text{Primary and Derived conditions on total dynamics} \\ \mathbb{Z}^*_\Omega(t_j) = \bigcup_{t=-\infty}^{t_j} \mathbb{Z}^*_\Omega(t_i) & \end{array}\right\}$$

(xi) There is the lower bound which is the primary category of varieties and there is no upper limit for derived varieties and hence the info-stock-flow is always at disequilibrium.

It is helpful to examine the information sets on the primary category of varieties and the information sets on the derived categories of varieties. The primary varieties constitute a column vector of the form $\mathscr{V} = [v_1(t_{-\infty}), v_2(t_{-\infty}), \cdots, v_i(t_{-\infty}), \cdots, v_\infty(t_{-\infty})]^T = [v_i(t_{-\infty}) | i \in \mathbb{I}^\infty]^T$ with a corresponding information stock which may be represented without the flow in the form $\mathscr{G} = \left[\mathbb{Z}^*_{v_1}(t_{-\infty}), \mathbb{Z}^*_{v_2}(t_{-\infty}), \ldots, \mathbb{Z}^*_i(t_{-\infty}), \ldots, \mathbb{Z}^*_{v_i}(t_{-\infty}), \ldots, \mathbb{Z}^*_{v_\infty}(t_{-\infty})\right]^T = \left[\mathbb{Z}^*_{v_i}(t_{-\infty}) | i \in \mathbb{I}^\infty\right]^T$. This primary category of varieties with the corresponding primary information stocks may be seen as initializing the beginning of times, where the total information stock which is available for informing, and knowing is $\mathbb{Z}^*_\Omega(t_{-\infty}) = \bigcup_{v \in \mathbb{V}} \mathbb{Z}^*_{v_2}(t_{-\infty})$. When the system of transformation begins, the information flows between time points are $\mathbb{Z}_v(t_{j+1}) = \mathbb{Z}^*_v(t_j) -, \mathbb{Z}^*_v(t_{j-1})$ for a discrete case and $\mathbb{Z}_v(t) = \frac{d\mathbb{Z}^*_v(t)}{dt}$ in continuous case.

The understanding of the complexity of the system reveals itself through the notion that the system is composed of primary category of infinite number of varieties the identities of which are established by the corresponding family of information stocks independent of the existence of other objects in the universe. The system, therefore, is a complex system of infinite structure of information relations. Each variety is under its own transformation process from within through the forces and energy that are internally generated to produce a variety-specific behavior which produces an information flow to continually update the info-stocks with the continual transformation of the information relation through infinite time structure without end. The nature of the information relation is such that the internal dynamics of each variety is affected by external conditions that indicate a framework of internal decision-choice response. It is here, that the concept of relational interactions between the external and internal environments find meaning and substance. In this respect, existence in general finds meaning in relations to things through the stock-flow process of information with continual solution-problem conflict. These relations to things through information generate problems the solutions of which generate new problems to establish a family of problem-solution processes with a changing structure of a family of families of information relations. The information relations find meanings in the family of inter-communication processes of the signal dispositions and with the corresponding methods of decoding which find interpretive actions in intra-communication processes to the characteristic dispositions.

Chapter 5
Info-dynamics: An Algebraic Introduction

Chapter 4 of this monograph provides variety-transformation dynamics on stock-flow information conditions on specific and the total system that have been made possible by technological set of transformations referred to as the set of $\mathfrak{T}-$Processes. It deals with two information processes $\mathbb{Z}-$Processes of specific and total varieties concerning the universal object set Ω, the set of phenomena Φ which together as a unit defines the variety space $\mathbb{V}$ as viewed from the space of characteristic-signal dispositions of the form $\mathbb{Z}{=}\mathbb{X} \otimes \mathbb{S}$, the collection of which constitutes the information space defined by $\mathbb{Z}^*_\Omega = \bigcup_{\nu\in\mathbb{V}} \bigcup_{t=+\infty}^{-\infty} \mathbb{Z}^*_\nu(t)$ of past, present and future of the universe of the form $\mathfrak{U} = (\Omega \otimes \Phi \otimes \mathbb{V} \otimes \mathbb{T}) = \{\mathfrak{u}{=}(\omega, \phi, \nu, t) | \omega \in \Omega, \phi \in \Phi, \nu \in \mathbb{V}, t \in \mathbb{T} = (-\infty, +\infty)\}$. The universal elements are seen as objects, phenomena, varieties in time and over time where time is neutral and may enter as an explicit or implicit variable. The two information processes of variety-specific dynamics and variety-total dynamics constitute an organic enveloping which is defined by a conditions of variety-matrix enveloping under an appropriate time structure.

The horizontal vectors represent individual stock-flow-variety dynamics over all time points. The column vectors represent sock-flow conditions of all varieties at any given time point. The time-variety matrix exists as an enveloping time-dated panel information stock-flow relations in complexity of systems dynamics. The systems dynamics may be automatic in the sense of internally induced by internally utilizing the necessary conditions to create the sufficient conditions required for internal-self transformation of the varieties. It may also be non-automatic in the sense of externally induced from without by externally utilizing the necessary conditions to create the sufficient conditions for external transformation of varieties. At the ontological space all transformations are automatic when the universal system is seen a unit with relational continuum with intersectionality induced by the existence of fuzzy categories. It has been pointed out that the difference between the internal process and external process is in transformation sustainability where the

K.K. Dompere, *The Theory of Info-Dynamics: Rational Foundations of Information-Knowledge Dynamics*, Studies in Systems, Decision and Control 114, https://doi.org/10.1007/978-3-319-63853-9_5

internal generated energy and the induced motion are continually sustainable while external inducement is always temporary and non-sustainable.

5.1 The Concepts and Structures of Information Process and Processor Under Socio-natural Technological Space

The transformations of varieties are *information processes* and generated by a set of *information processors* to change the set of *information relations* in the universal object set and the corresponding phenomena. The changes in the information relations among ontological objects under decision-choice processes act to create new information flows to augment the existing info-stock. The information processors are induced by technologies form the technological space $\mathfrak{T}$. The cardinality of the set of information processors is equal to the cardinality of the varieties at any time point. The set of the known varieties is a set of knowledge processes and generated by a set of *knowledge processors*. Over the epistemological space, these knowledge processors create social technologies while the social technologies create the knowledge processors. The cardinality of the set of knowledge processors is a small subset of the set of information processors. There are two initial important concepts of process and processor in all systems as seen from a set of information relations. The process and the processor are related to time. Thus given the time set $\mathbb{T} = (-\infty, +\infty)$ and the primary variety-set $\mathbb{V} = \{v_i | i \in \mathbb{I}^{\infty}\}$, an epistemic space is open to define the concepts of a process and a processor. In the study of $\mathfrak{T}-$Processes and the corresponding $\mathbb{Z}-$Processes, It may be noted that the primary variety set may be arbitrarily selected with a define time point under the principle of ancestor-successor (or parent-offspring) connectivity, where the path to ancestry is a leftward telescopic view through the methodological reductionism and the path to successors (offspring) is a rightward telescopic view through the methodological constructionism given the initial time point to establish the past-present-future time connectivity under the *Sankofa-anoma* conditions.

Definition 5.1.1 (*Information Process*)
Let $\mathbb{T} = (-\infty, +\infty)$ be the time set and $\mathbb{V} \otimes \mathbb{T} = \{(v, t) | v \in \mathbb{V}, t \in \mathbb{T}\}$ be the variety set over time, then $\mathscr{P}_v$ is an information process if it changes the time-point characteristic disposition of the same variety of itself from in $\mathscr{P}(v, t_i) \to (v, t_j)$ such that $\mathbb{Z}_v^*(t_i) \subset \mathbb{Z}_v^*(t_j)$ in a discrete time case and in continuous time case $\dot{\mathbb{Z}}_v^*(t) = \frac{d\mathbb{Z}_v^*}{dt} > 0$ with variety difference of the form $(v, t_i) \not\approx (v, t_j)$. The collection of all the processes on $\mathbb{V} = \{v_i | i \in \mathbb{I}^{\infty}\}$ over time $\mathbb{T}$ is a family of family of processes of the form $\mathscr{P} = \{\mathscr{P}_{v_i}(t) | i \in \mathbb{I}^{\infty}, v \in \mathbb{V} \,\&\, t \in \mathbb{T}\}$. The information processor $\mathscr{P}_v$ is a technological processors from the technological space $\mathfrak{T}_v$ where $\mathscr{P}$ is a set in the technological space $\mathfrak{T}$. In other words, the information process and

information processor are variety transformation process and processors respectively.

Every object $\omega \in \Omega$ at any given time is also a variety $\nu \in \mathbb{V}$ and a phenomenon $\phi \in \Phi$ with an identity given by its information structure $\mathbb{Z}_\nu$. The time path of every variety is connected by an information process such that there are pre-successor and a successor defined by transformations over the time domain. The pre-successor is an input into the transformation process of the information relation leading to an output which then becomes an input. The pre-successors are the *ancestors*, the successors are the *offspring* and the future successors are the future info-generation such that each variety is an infinitely time dated vector that may be represented with its information process as $\mathbb{V} \otimes \mathbb{T} \otimes \mathscr{P} = \{(\nu, t, \mathscr{P}_\nu) | \nu \in \mathbb{V}, t \in \mathbb{T}, \mathscr{P} \in \mathscr{P}\}$. Every information process is also an information processor with the ancestor-successor structure as an input-output information relation. Every time dated variety is simultaneously an input and output. This input-output information relation is basically an ancestor-successor relation, where every variety is an output from ancestor as well as serving as an input for a successor. Given the time set $\mathbb{T}$, the variety space $\mathbb{V}$ and the space of processors $\mathscr{P}$ the enveloping of each variety and the enveloping space may be defined. The information process as used here is different from the tradition. It is the creation and destruction of varieties with a continual expansion of info-stock on individual or collective varieties. The individual and collective varieties include all socio-natural varieties. In other words, the info-stocks and info-flows include ontological information and epistemological information.

Definition 5.1.2 (*Information Enveloping Path and Space*)

The *information enveloping* $\mathfrak{E}(t)$ of each variety $\nu \in \mathbb{V}$ is the time path of variety transformations of the same primary variety equipped with different time-dependent information processors $\mathscr{P}_\nu \in \mathscr{P}$ of, collection of differential time-dependent varieties $\nu(t) \in (\mathbb{V}, \mathbb{T})$, of information stock and variety processor with respect to a particular variety such that $\mathfrak{E}_\nu(t) = \{(\nu(t), \mathscr{P}_\nu(t), \mathbb{Z}_\nu^*(t)) |$, $\nu \in \mathbb{V}, \mathscr{P}_\nu \in \mathscr{P}, t \in \mathbb{T}, \&, \ \mathbb{Z}_\nu^* = (\mathbb{X}_\nu^* \otimes \mathbb{S}_\nu^*)\}$. The collection of all the elements of the individual variety enveloping constitute the enveloping space of the form $\mathscr{E} = \{\mathfrak{E}_\nu(t) | \forall \nu \in \mathbb{V} \, \& \, \forall t \in \mathbb{T}, \}$ in the information space $\mathbb{Z}_\Omega^* = \bigcup_{\nu \in \mathbb{V}} \bigcup_{t=+\infty}^{-\infty} \mathbb{Z}_\nu^*(t)$.

The information enveloping $\mathfrak{E}_\nu(\cdot)$ for each variety and the enveloping space $\mathscr{E}(\cdot)$ for all varieties are in relation to the nature of the complex transformation processes of socio-natural *decision-choice system* and the structure of management of associated command-control system, collectively or individually defined. The management of the command-control system of each information processor may be an internal process induced from internal conflicts in the internal duality that produce energy for the work of change through an internal continual motion. The conflicts of the internal duality with relational continuum is manufactured by internal decision-choice technology. Here, the information generating activities of ancestor-successor processes find linguistic expressions in the structures of self-organizing, automatic-control, self-exciting, self-correcting, self-learning, self-adapting and many others, where variety self-motion is the result of the behavior of these structures. The management of the command-control system of an information processor may be an external process which depends on externally produced energy for motion and change. The internal process generates internal energies that produce sustainable transformations of varieties while the external process does not produce sustainable transformations. For example, a car does not produce itself while seeds among a multitude of objects in the universal object set generally produce themselves.

In fact, the whole engineering process of automation is human attempts to mimic the internal behavior of nature with energy constraints. The information processes and processors are translated into cost-benefit information duality with relational continuum and unity for each variety in decision-choice activities of the transformation process with cost-benefit rationality under the *asantrofi-anoma principle* where one cannot select real benefit and leave the real cost behind. Here, it may be keep in mind that every variety sits on real cost-benefit duality with a relational continuum and unity, where the comparative analytics for transformations must be seen in terms of net benefit, where the net benefit of the ancestor is compared with the net benefit of a successor and similarly, the cost of the ancestor is compared with the cost of the successor. In terms of economics analytics, the real cost-benefit relationality is defined in terms of opportunity cost where the real cost of acquiring the real benefit of the successor is the benefit forgone of the pre-successor [R5.13], [R5.14]. Every pre-successor is an actual variety which belongs to the space of reality, $\mathfrak{A}$ and every successor is a potential variety which belongs to the space of the potential $\mathfrak{U}$ and all the ancestors have available information as part of the info-stocks as well as belonging to the potential space $\mathfrak{U}$ as seen in the space of actual-potential polarities with transformation dynamics. As has been explained over the ontological space the potential is possible and hence there is no need to go over the possibility space in transformation ($\mathfrak{U} = \mathfrak{P}$). Here, the structure of information relations is defined as the structure of cost-benefit relations which is the structure of input-output relations that translate into the structure of pre-successor-successor relations. The enveloping of any variety is a path of historic past, historic present and historic future in relational continuum and unity, where the path is a stock-flow information process, for the variety's ancestry and offspring that allows the development of scientific and non-scientific knowledge types by cognitive agents over the epistemological space.

Given the defined pure time set, the information systems dynamics is, thus, governed by either an internal, external or both organic decision-choice system on the basis of intentionalities that are linked to input-output conditions of costs and benefits of the socio-natural transformations that follow cost-benefit rationality under the *asantrofi-anoma* principle. Let us recall that every transformation of a variety is a socio-natural decision-choice info-process that involves inputs and outputs where inputs find expression in cost-information and outputs find expression in benefit-information. This is the cost-benefit variety decision-choice structure contained in the *asantrofi-anoma principle* which simply presents the notion that every decision-choice item or a variety resides in a cost-benefit duality with relational continuum and unity, where the cost and benefit mutually create themselves in an inseparable decision-choice unit such that the benefit a decision-choice action cannot be selected without the cost. The implication, here, is that every decision-choice action on a variety contains cost and benefit which are mutually supporting. The *asantrofi-anoma* principle says the cost and benefit are inseparable in every decision-choice action on varieties. The cost-benefit rationality say that choose a variety that has the highest benefit among all possible varieties. In fact, this is the foundation of cost-benefit analysis and cost-benefit rationality where the benefits and costs of each transformation decision are compared for advantage before a change of a variety.

The time set $\mathbb{T}$ and the set of primary varieties $\mathbb{V}$ create an extensive column vector at each point of time and a complex matrix of derived outcomes of an enveloping, where each column vector at each point of time is a derived vector of varieties as well as serving as a primary vector or a vector of pre-successors for the next time period vector of successors creating an infinite structure of an evolving matrix of varieties with a corresponding matrix of information info-stocks on varieties and a corresponding matrix of info-flows. The transformation dynamics involves three important matrixes of the matrix of varieties, the matrix of info-stocks and the matrix of info-flows, where the stock-flow matrixes allow the establishment of a matrix variety identities. The evolving variety matrix is called *variety transformation transitional matrix* (VTTM) and is of the form $\mathfrak{W} = (\mathbb{V} \otimes \mathbb{T})$ where both elements in $(\mathbb{V} \otimes \mathbb{T})$ are infinite vectors of varieties and time. Corresponding to the variety transformation transitional matrix is the *information transformation transitional matrix* and *information processor transitional matrix*.

It is useful to examine the information sets on the *primary category of varieties* and the information sets on the *derived categories of varieties* through a dual *telescopic processes* of constructionism-reductionism dualistic time relation.

Proposition 5.1.1 (Forward Composition of Processors)

Let $\mathscr{P} = \{\mathscr{P}_{v_i}(t) | i \in \mathbb{I}^{\infty}, v \in \mathbb{V} \,\&\, t \in \mathbb{T}\}$ *be the space of processors and corresponding to it an information space* $\mathbb{Z}_{\Omega}^{*} = \bigcup_{v \in \mathbb{V}} \bigcup_{t=+\infty}^{-\infty} \mathbb{Z}_{v}^{*}(t)$ *then for* $\forall \mathscr{P}_{v_i}(t) \in \mathscr{P}$, *it is the case that* $\mathscr{P}_{v_i}(t_i) \neq \mathscr{P}_{v_i}(t_j)$ *and* $\mathbb{Z}_{v}^{*}(t_i), \mathbb{Z}_{v}^{*}(t_j) \in \mathbb{Z}_{\Omega}^{*}$ *is such that* $\mathbb{Z}_{v}^{*}(t_i) \subset \mathbb{Z}_{v}^{*}(t_j), \forall (t_j - t_i) > 0 \in \mathbb{T}$ *and that it is the case that* $\mathscr{P}_{v_i}(t_j) \circ (\mathscr{P}_{v_i}(t_i)) \Rightarrow \mathscr{P}_{v_i}(t_i) \not\phi \mathscr{P}_{v_i}(t_j), \forall i, j \in \mathbb{N}^{+}, (t_j - t_i) > \in \mathbb{T}$. *The symbol* $(\circ)$ *is a composition of processors acting on the same variety over infinite time and the symbol* $(\not\phi)$ *is irreversibility. The processes are forward transformation and right telescopic into historic future from historic present about information on future varieties, where the processors are constructionist in nature into the infinite future.*

Proposition 5.1.2 (Backward Composition of Processors)

Let $\mathscr{P} = \{\mathscr{P}_{v_i}(t) | i \in \mathbb{I}^{\infty}, v \in \mathbb{V} \,\&\, t \in \mathbb{T}\}$ *be the space of processors and corresponding to it an information space* $\mathbb{Z}_{\Omega}^{*} = \bigcup_{v \in \mathbb{V}} \bigcup_{t=+\infty}^{-\infty} \mathbb{Z}_{v}^{*}(t)$ *then for* $\forall \mathscr{P}_{v_i}(t) \in \mathscr{P}$, *it is the case that* $\mathscr{P}_{v_i}(t_i) \neq \mathscr{P}_{v_i}(t_j)$ *and* $(\mathbb{Z}_{v}^{*}(t_i), \mathbb{Z}_{v}^{*}(t_j) \in \mathbb{Z}_{\Omega}^{*})$ *is such* *that* $\mathbb{Z}_{\mathbb{V}}^{*}(t_i) \subset \mathbb{Z}_{\mathbb{V}}^{*}(t_j), \forall (t_i - t_j) = 0$ *and that it is the case that* $\mathscr{P}_{\mathbb{V}_i}(t_j) \circ (\mathscr{P}_{\mathbb{V}_i}(t_i)) \Rightarrow \mathscr{P}_{\mathbb{V}_i}(t_i) \circ \mathscr{P}_{\mathbb{V}_i}(t_j), \forall i,j \in \mathbb{N}^{+}, \ (t_j - t_i) > 0 \in \mathbb{T}$ then $\exists f(\cdot) \ni f(\mathscr{P}_{\mathbb{V}_i}(t_j)) = \mathscr{P}_{\mathbb{V}_i}(t_i)$ *where* $\mathscr{P}^{-}_{\mathbb{V}_i}(t_j) = \mathscr{P}_{\mathbb{V}_i}(t_i)$. *The processes are backward transformations which are left telescopic into the infinite historic past from historic present and the processors are reductionists in nature over information on actual and perhaps potential varieties towards infinite past. The importance of the constructionist and reductionist info-processors and processes is that one can trace the evolution of the primary varieties in terms of the derived varieties in the forward process, and one can trace the path of the derived varieties in a backward process to the primary varieties in the forms of ancestral-offspring structural dynamics from the specified present or defined set of initial conditions as shown in* Fig. 5.1.

Figure 5.1 has some interesting ideas. It has a process connectivity in relational continuum and unity. The continuity is of two types of present-future connectivity that asserts no limitations on progress of socio-natural transformations of varieties on ancestor-successor or *parent-offspring* process into the infinite future and present-past connectivity that reveals no limitation on acquaintances, informing, learning and the history of past transformations of varieties on successor-ancestor or *offspring-parent* process into the infinite past. This simply means that for every known and unknown actual-potential successor existence, there is an ancestor which may be viewed as the primary variety. The variety space $\mathbb{V}$ is under the matter-energy processes and the information space $\mathbb{Z}_{\Omega}^{*}$ is under socio-natural decision-choice processes that are connected to the matter-energy processes. It must be clear that the natural decision-choice processes that are connected to natural transformations are independent of cognitive agents who may be completely ignorant to the existence of the processes and outcomes.

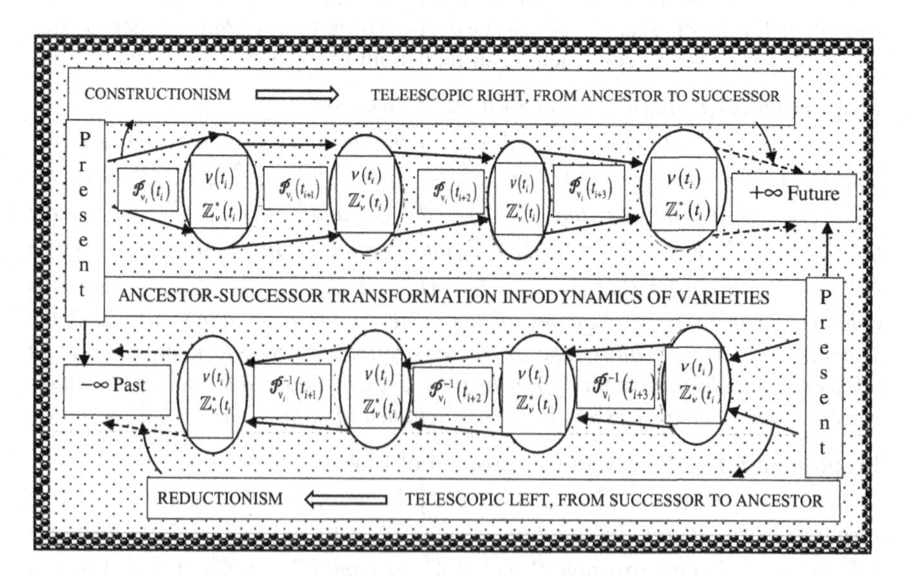

Fig. 5.1 Telescopic picture of ancestral future, ancestral present and ancestral past with transformations induced by socio-natural technological processors

The social decision-choice processes perform two important tasks of limited transformation on society and the creation of knowledge about the universal structure of socio-natural transformations. To reveal the inner structure of relationality of continuum and unity between the set of matter-energy processes and the set of decision-choice processes in continual information production, let us take a look at the initial variety vector and the time processes in periods of transformation over the infinite times $\mathbb{T}$ in the form $(\mathbb{V}_{t-\infty} \otimes \mathbb{T}) = \{(\mathrm{v}_i, t_{-\infty}) | i \in \mathbb{I}^{\infty}, t_{-\infty} \in \mathbb{T} = (-\infty, +\infty)\}$ which initializes the beginning. In the current analytical framework, info-statics is the study of the meaning and distinction among socio-natural varieties, where the theory of info-statics is concerned with a search for principles that can accurately describe the information content, storage and its communication among ontological elements in a given time point in the universal order allow categories to be form. The theory of info-dynamics, on the other hand, is the study of information flows under transformations of matter-energy structures. It is thus concerned with the search for dynamic principles that can be used to accurately describe the stock-flow behavior of information production at the level of particular and general socio-natural varieties and categorial varieties. In simplicity, it is the study of transformations and changes induced by qualitative and quantitative motions under elements from the technological space.

5.2 The Concepts of Identification Transformation-Transitional Variety Matrix with Primary and Derived Categorial Varieties in a Simple Algebraic Structure

Each initial variety has a transformation time path that is a row vector in the time process. Every variety in the same category is under the same conditions of dynamic behavior. In terms of the transformation-transitional variety matrix, an examination shall be conducted on column vector corresponding to the behavior of the case of column transformation where an element in a row changes its characteristics and then migrates into a different row.

Definition 5.2.1 (*Variety-Matrix Transformation Relation Matrix (VTRM)*)
A variety-matrix transformation relation, $\mathfrak{T}$ is such that given a variety set $\mathbb{V}$ and a time set $\mathbb{T}$, $\mathbb{V} = \{\mathrm{v}_{-\infty} \ldots \mathrm{v}_{-1}, \mathrm{v}_0, \mathrm{v}_1, \mathrm{v}_2, \ldots, \mathrm{v}_{+\infty}\}$ and $\mathbb{T} = \{t_{-\infty} \ldots t_{-1}, 0, t_1, t_2 \ldots, t_{+\infty}\}$, $\mathfrak{T} \subset \mathbb{V} \otimes \mathbb{T}$ we may define $\mathfrak{T} = \{\mathbb{V} \otimes \mathbb{T} | \mathfrak{w}_{\mathfrak{T}}(\mathrm{v}, t) = \mathfrak{w}$ where $\mathrm{v}, \mathfrak{w} \in \mathbb{V}, \mathfrak{w} = \mathscr{T}(\mathrm{v})$ and $t \in \mathbb{T}\}$, where $\mathfrak{w}$'s $\not\approx$ v's belong to the set of transformed varieties from the $(\mathrm{v} - t)$ cohort objects and $\mho_{\mathfrak{T}}$ is a matrix of the transformed varieties with *transformation processor operators* $\mathfrak{w} = \mathscr{T}(\mathrm{v})$ which is infinitely defined, where $\mathfrak{w}$'s are the successors or offspring and v's are the ancestors or parents in the transformation process of categorial varieties.

Definition 5.2.2 (*Variety Transformation Transitional Matrix*)
Given a variety-matrix identification transformation, $\mathfrak{T}$, the time-enveloping of transformation variety-matrix, $\mho_{\mathfrak{T}}$ is defined by a $(t_{-\infty}, t_{+\infty})$ cohort of transformed relational varieties where there are infinite-roles of defined varieties of the same elements and infinite-columns of other defined varieties of different elements and the matrix $\mathfrak{w} = \mho_{\mathfrak{T}}$ is the *variety transformation transitional matrix* (VTTM) that provides the characteristics of the object in each cohort.

Definitions 5.2.1 and 5.2.2 may be represented as in Fig. 5.2.

In the matrix $\mho_{\mathfrak{T}}$ $t_j = j$th time in the transformation process and $v_i = i$th variety. The table may be written as a transformation transitional matrix of the form:

$$\mho_{\mathfrak{T}}(\cdot) = \begin{pmatrix} \mathfrak{w}_{1,t_{-\infty}} & \cdots & \mathfrak{w}_{1,t_{+\infty}} \\ \vdots & \mathfrak{w}_{i,t_j} & \vdots \\ \mathfrak{w}_{+\infty,t_{-\infty}} & \cdots & \mathfrak{w}_{+\infty,t_{+\infty}} \end{pmatrix} = \left[\mathfrak{w}_{i,t_j}\right] \tag{5.2.1}$$

The vector $\mathbb{V} = \left[v_1 \ldots v_2, \ldots v_j \ldots, v_{+\infty}\right]^T$ is called a *socio-natural primary vector*, which constitutes the primary varieties in an arbitrary initialized time such as $t = -\infty$. The primary varieties belong to the primary category and form the initial inputs into the transformation processing operators $\mathcal{T}(\cdot) \in \mathfrak{T}$. They are the initial ancestors or parents from which offsprings or successors are derivable by constructionism or to which offspring or successors are traceable by reductionism. Corresponding to the primary varieties is the primary vector of information stock conditions of the form $\mathbb{Z}_{\mathbb{V}}^* = \left[\mathbb{Z}_{1,t_{-\infty}}^* \ldots \mathbb{Z}_{2,t_{-\infty}}^*, \ldots \mathbb{Z}_{j,t_{\infty}}^* \ldots, \mathbb{Z}_{+\infty,t_{-\infty}}^*\right]^T$. The transformation processing operators are also infinite where each primary variety has

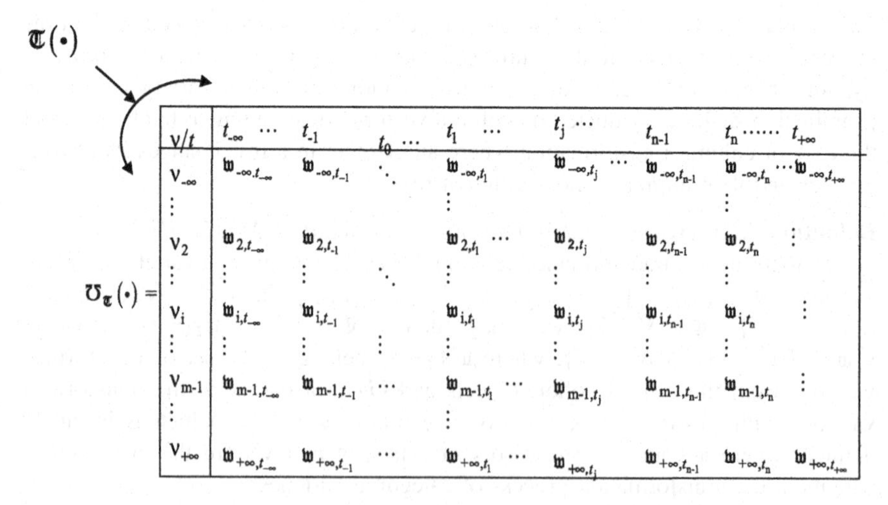

Fig. 5.2 A conceptual structure of an identification transformation-transitional relational matrix of varieties in an infinite time $\mathbb{T} = (-\infty, +\infty)$

infinite horizontal vector of transformation-processing operators $\mathfrak{T}_{\mathrm{i}} \subset \mathfrak{T}$ and where the (i) indicates the row vector. Each *i*th-row-transformation processor is of the form $\mathfrak{T}_{\mathrm{i}} = \langle \mathcal{T}_{\ell t_j}(\cdot) \mid \ell \in \mathbb{I}_{C}^{\infty}, t_j \in \mathbb{T}, i, j \in \mathbb{N} \rangle$ where ℓ runs across the columns with any fixed *i*th-row and $\mathfrak{T} = \langle \mathfrak{T}_{\mathrm{i}} \mid \mathrm{i} \in \mathbb{I}^{\infty} \rangle$. The collection of the infinite *transformation processing operators* is a matrix of the form $\mathfrak{T}(\cdot) = [\mathcal{T}_{\ell t_j}(\cdot)]^{\infty}$ called *transformation-transitional identification processing matrix*. The vector $\mathbb{T} = \{t_{-\infty}, \ldots t_0, \ldots t_j \ldots, t_{+\infty}\}$ is called a *transformation-time vector* where $\mho_{\mathfrak{T}}(\cdot) = [\mathfrak{w}_{\mathrm{i},t_j}]^{\infty}$ is the *transition variety-identification matrix*. Corresponding to the variety identification matrix are the *transformation object identification matrix*, of the form $\mho_{\Omega}(\cdot) = [\omega_{\mathrm{i},t_j}]^{\infty}$, *transformation phenomenon identification matrix* of the form $\mho_{\Phi}(\cdot) = [\phi_{\mathrm{i},t_j}]^{\infty}$ and *transformation information identification matrix* of the form $\mho_{\mathbb{Z}^*}(\cdot) = [\mathbb{Z}^*_{\mathrm{i},t_j}]^{\infty}$. The transition variety identification matrix is composed of *derived elements* under the principle of substitution-transformation duality which is made possible by the transformation information identification matrix that points to the structures of the *transformation object identification matrix* which is the evolving identity matrix which is obtained from the corresponding *transformation phenomenon identification matrix* viewed in the methodological constructionism-reductionism system of thought. The transformation-identification matrix is generated by transformation-processing operator matrix. The methodological constructionism allows the creation of the derived categorial varieties while the methodological reductionism allows the identification of primary categories of the derived varieties. It is useful to keep in mind that the primary category is the source and the mother of derived categories and that corresponding to the transformation transitional matrix is also decision-choice processing matrix of the form $\mho_{\mathfrak{D}}(\cdot) = [\mathcal{D}_{\ell t_j}(\cdot)]^{\infty}$.

5.2.1 *Transformation-Transitional Matrix Analytics of Varieties*

It is now useful to conduct analytics of the relational contents of the essential matrixes that present conditions to construct the theory of info-dynamics and examine the underlying general principles. The essential matrixes are summarized in Table 5.1. There are two primary vectors and six active matrixes in interdependent actively operating systems that generate derived categorial varieties and stock-flow info-dynamics where,

$$\mho_{\mathbb{Z}^*}(\cdot) = \begin{pmatrix} \mathbb{Z}^*_{1.t_{-\infty}} & \cdots & \mathbb{Z}^*_{1,t_{+\infty}} \\ \vdots & \mathbb{Z}^*_{\mathrm{i},t_j} & \vdots \\ \mathbb{Z}^*_{+\infty,t_{-\infty}} & \cdots & \mathbb{Z}^*_{+\infty,t_{+\infty}} \end{pmatrix} = [\mathbb{Z}^*_{\mathrm{i},t_j}]^{\infty} \quad \text{(Info-stock)} \qquad (5.2.1.1)$$

Table 5.1 A summary of essential quantitative and qualitative transformation dynamics with categories of quantitative and qualitative motion in vectors and matrixes

Socio-natural primary categorial varieties	$\mathbb{V} = [v_1 \ldots v_2, \ldots v_i \ldots, v_{+\infty}]^T$	(1)
Transitional time vector	$\mathbb{T} = [t_{-\infty}, \ldots t_0, \ldots t_j \ldots, t_{+\infty}]$	(2)
Transformation processing operator matrix	$\mathfrak{T}(\cdot) = [\mathscr{T}_{\ell t_j}(\cdot)]^{\infty}$	(3)
Transformation info-flow identification matrix	$\mho_{\mathbb{Z}}(\cdot) = [\mathbb{Z}_{i,t_j}]^{\infty}$	(4)
Transformation info-flow identification matrix	$\mho_{\mathbb{Z}^*}(\cdot) = [\mathbb{Z}^*_{i,t_j}]^{\infty}$	(5)
Transition variety-identification derived matrix	$\mho_{\mathfrak{T}}(\cdot) = [\mathfrak{w}_{i,t_j}]^{\infty}$	(6)
Transformation object identification matrix	$\mho_{\Omega}(\cdot) = [\omega_{i,t_j}]^{\infty}$	(7)
Transformation phenomenon identification matrix	$\mho_{\Phi}(\cdot) = [\phi_{i,t_j}]^{\infty}$	(8)
Decision-choice processing matrice	$\mho_{\mathfrak{D}}(\cdot) = [\mathscr{D}_{\ell t_j}(\cdot)]^{\infty}$	(9)
Net cost-benefit relational matrix	$\mho_{\mathbb{B}}(\cdot) = [\mathrm{B}_{\ell t_j}(\cdot)]^{\infty}$	(10)

Where $\mathbb{I}_{\mathbb{V}}^{\infty}$ = Index set of categorial varieties with a generic element i, $\ell \in \mathbb{I}_{\mathbb{V}}^{\infty}$ and $\mathbb{I}_{\mathbb{T}}^{\infty}$ = Index set of time positions with a generic element of $j \in \mathbb{I}_{\mathbb{T}}^{\infty}$

$$\mathfrak{T}(\cdot) = \begin{pmatrix} \mathscr{T}_{1.t_{-\infty}} & \cdots & \mathscr{T}_{1,t_{+\infty}} \\ \vdots & \mathscr{T}_{i,t_j} & \vdots \\ \mathscr{T}_{+\infty,t_{-\infty}} & \cdots & \mathscr{T}_{+\infty,t_{+\infty}} \end{pmatrix} \tag{5.2.1.2}$$

$$= [\mathscr{T}_{i,t_j}]^{\infty} \text{(Transformation-Operator Matrix)}$$

$$\mho_{\Omega}(\cdot) = \begin{pmatrix} \omega_{1.t_{-\infty}} & \cdots & \omega_{1,t_{+\infty}} \\ \vdots & \omega_{i,t_j} & \vdots \\ \omega_{+\infty,t_{-\infty}} & \cdots & \omega_{+\infty,t_{+\infty}} \end{pmatrix} = [\omega_{i,t_j}]^{\infty} \text{(Object Universe)} \tag{5.2.1.3}$$

$$\mho_{\Phi}(\cdot) = \begin{pmatrix} \phi_{1,t_{-\infty}} & \cdots & \phi_{1,t_{+\infty}} \\ \vdots & \phi_{i,t_j} & \vdots \\ \phi_{+\infty,t_{-\infty}} & \cdots & \phi_{+\infty,t_{+\infty}} \end{pmatrix} \tag{5.2.1.4}$$

$$= [\phi_{i,t_j}]^{\infty} \text{(The Phenomenon Universe)}$$

$$\mho_{\mathfrak{D}}(\cdot) = \begin{pmatrix} \mathcal{D}_{1,t_{-\infty}} & \cdots & \mathcal{D}_{1,t_{+\infty}} \\ \vdots & \mathcal{D}_{i,t_j} & \vdots \\ \mathcal{D}_{+\infty,t_{-\infty}} & \cdots & \mathcal{D}_{+\infty,t_{+\infty}} \end{pmatrix} \tag{5.2.1.5}$$

$$= \left[\mathcal{D}_{i,t_j}\right]^{\infty} \text{(The Decision-choice Universe)}$$

$$\mho_{\mathbb{B}}(\cdot) = \begin{pmatrix} \mathrm{B}_{1,t_{-\infty}} & \cdots & \mathrm{B}_{1,t_{+\infty}} \\ \vdots & \mathrm{B}_{i,t_j} & \vdots \\ \mathrm{B}_{+\infty,t_{-\infty}} & \cdots & \mathrm{B}_{+\infty,t_{+\infty}} \end{pmatrix} \tag{5.2.1.6}$$

$$= \left[\mathrm{B}_{i,t_j}\right]^{\infty} \text{(Relative Net-Benefit Universe)}$$

These matrixes may be summarized in a tabular form with corresponding role names for quick analytical reference in Table 5.1. On the basis of the behavior of the elements in the matrixes, basic principles of info-dynamics are the presented as laws of info-dynamics that allow distinctions among matter and energy. The transformation operators are also the processors that act on the corresponding varieties.

The set of the summary conditions in Table 5.1 presents a complex system of socio-natural transformational dynamics. It is the existence of these conditions that a complexity theory acquires meaning and analytical relevance towards the *concept of real definition* and the knowledge production system over the epistemological space. The theory of info-dynamics that is sought here is about the stock-flow dynamic behavior of $\left\{\mho_{\mathbb{Z}^*}(\cdot) = \left[\mathbb{Z}^*_{i,t_j}\right]^{\infty} \text{ and } \mho_{\mathbb{Z}}(\cdot) = \left[\mathbb{Z}_{i,t_j}\right]^{\infty}\right\}$ of information process that creates conditions of informing, knowing, learning communicating and teaching. The main attention in this monograph will center on how the transformation-transitional information matrix is manufactured from within the socio-natural systems. Corresponding to the matrixes $\left\{\mho_{\mathbb{Z}^*}(\cdot) = \left[\mathbb{Z}^*_{i,t_j}\right]^{\infty}\right.$ and $\left.\mho_{\mathbb{Z}}(\cdot) = \left[\mathbb{Z}_{i,t_j}\right]^{\infty}\right\}$, there are two infinite vectors $\mathbb{V}$ and $\mathbb{T}$. Given the two infinite vectors, there are four matrixes. Any fixed $t_j \in \mathbb{T}$ locates a column which defines info-statics and the corresponding theory of info-statics which includes the study of conditions of real definition of elements, varieties (species) and categorial varieties, identities and difference in relation to nominalism, vocabulary and language. Similarly, any fixed $v_i \in \mathbb{V}$ locates a row which specifies info-dynamics in the sense that $v_i \in \mathbb{V}$ is a primary categorial element from which derived categories $\mathfrak{w}_{i,t_j} \in \mho_{\mathfrak{T}}(\cdot)$ are transformed by the transformation processing operators, $\mathcal{T}_{\ell t_j} \in \mathfrak{T}(\cdot)$. The net cost-befit transitional matrix is extremely important in the epistemological space, where decision-choice processes involves subjective judgments in terms of relative cost-benefit comparisons of potential varieties viewed individually by cognitive agents under the cost-benefit rationality in terms of a replacement or a non-replacement of the existing variety.

5.2.1.1 The Transformation-Transitional Matrix, Categorial Conversion and Necessity

The structure of universal info-dynamics must be seen in two sub-structures of the ontological info-dynamics and the epistemological info-dynamics as organic varieties with organic identities in relational continuum and unity. These two structures must be kept in mind when one views the matrixes which represent the general and the specific info-production systems. The study of the information structure on the column vector $\mathbb{V}$ as the primary category and any column vector taken as defining the conditions of subsequent evolution presents the necessary conditions for transformation. The theory designed for the study of these necessary conditions for transformation in destruction-construction duality is called the *theory of categorial conversion* [R17.15]. The theory of categorial conversion presents the necessary conditions of information on any element in the primary column vector to be internally transformed from the initial categorial identity. This necessary conditions define the socio-natural necessity in the dynamic space of actual-potential polarities which contains the sub-spaces of negative and positive dualities with relational continuum and unity. The concepts of relational continuum and unity have been explained. The relational continuum presents a situation where every negative characteristic set has a positive-set characteristic support while the relational unity affirms unified conditions that define the identity of the duality.

From the principle of categorial conversion, any element $\mathfrak{w}_{\mathrm{it_j}}$ and hence corresponding information of any of the column of the $\mathrm{t_j}$-column vectors may be considered as a primary category from which the subsequent categorial elements of $\mathfrak{w}_{\mathrm{it_\ell}}$ may be derived where $\left((\mathrm{t_\ell}\text{-}\mathrm{t_j} > 0) \in \mathbb{T}\right)\left(\mathbb{Z}^*_{\mathrm{it_j}} \cap \mathbb{Z}^*_{\mathrm{it_\ell}}\right) = \mathbb{Z}^*_{\mathrm{it_j}}$. The last statement is an expression of the principle of non-uniqueness of primary category [R17.15], [R17.16]. The principle of non-uniqueness of the primary category in the study of transformation or any dynamics defines a powerful condition for transformations, where the initial conditions for the study of destruction-construction processes or space-time processes are not unique and may be arbitrarily selected. The necessary conditions establish the command and control instruments that must be manufactured under the principle of need, freedom and necessity. The command and control instruments must be manufacture by a decision-choice system. The successful instrumentation makes available to the decision-choice system the necessary conditions of the management of the command and control elements. Let such necessary conditions be specified as $\mathfrak{N}(\cdot)$ which will vary from variety to variety and from column vector to column vector as specified in the transformation-transitional-processing matrix and the summary in Table 5.1. Let an attention be turned now to the creation of sufficient conditions such that $\mathrm{v_i} \in \mathbb{V}$ with information

$\mathbb{Z}_{\mathrm{i}}^{*}$ will change its variety to $\mathfrak{w}_{\mathrm{it_j}} \in \mho_{\mathfrak{T}}(\cdot)$ with information $\mathbb{Z}_{\mathrm{it_j}}^{*}$ where $\mathfrak{w}_{\mathrm{it_j}} \not\approx \mathrm{v_i}$ and $\mathbb{Z}_{\mathrm{i}}^{*} \subset \mathbb{Z}_{\mathrm{it_j}}^{*}$ for all i's as specified under categorial conversion.

5.2.1.2 The Transformation-Transitional Matrix and Decision Under Philosophical Consciencism

The theory of categorial conversion is used to study the necessary conditions for information expansion through categorial transformation of varieties where each variety is uniquely define by the categorial information set. It is devoted to the study of the conditions of info-statics, information stock of variety identification and the specification of initial conditions of info-dynamics at any time point in the sense of continual transformations of varieties and categorial varieties. The necessary conditions are not sufficient to bring about transformations of varieties and the updating of the information stocks. Let us keep in mind that for each given $\mathfrak{w}_{\mathrm{it_j}} \not\approx \mathrm{v_i}$ the equation of motion from $\mathrm{v_i} \in \mathbb{V}$ to $\mathfrak{w}_{\mathrm{it_j}} \in \mathbb{V}$ is revealed by the symmetric info-set difference $\left(\mathbb{Z}_{\mathrm{it_j}}^{*} \nabla \mathbb{Z}_{\mathrm{i}}^{*}\right)$ when the sufficient conditions are met. The *symmetric info-set difference* may be studied by combining it with the *info-set relative complement* $\left(\frac{\mathbb{Z}_{\mathrm{it_j}}^{*}}{\mathbb{Z}_{\mathrm{i}}^{*}}\right)$ in addition to info-set containment. It may be kept in mind that the structure of the info-dynamics is conic stack that contains the initial cone and conically expends outward in a time-forward motion. It also conically reduces inward in a time-backward motion to the initial cone. The study of the sufficient conditions containing freedom is the study of transformation-transitional processing operators $\mathcal{T}_{\ell t_j} \in \mathfrak{T}(\cdot)$, where each of these transformation processing operators is generated by a decision-choice system with or without intentionality. The conditions of non-intentionality may apply to natural transformations of varieties and categorial varieties while the conditions of intentionality may apply to social transformations of varieties and categorial varieties.

5.3 Information Production and the Principle of Opposites

From the view point of sustainability of information generation, our main concern will be concentrated on the internally induced transformations. The theory of info-dynamics as reflecting the internal transformation process will then be related to conditions of externally imposed changes of any given variety. The internal transformation dynamic process must be shown to be induced from within in terms of how the existing characteristic-signal disposition of say a variety $\left(\mathbb{Z}_{\ell}^{*}(t_1), \mathrm{v}_{\ell}(t_1)\right)$ changes to a new characteristic-signal disposition with different variety $\left(\mathbb{Z}_{\ell}^{*}(t_2), \mathrm{v}_{\ell}(t_2)\right)$ such that $\mathbb{Z}_{\ell}^{*}(t_1) \not\approx \mathbb{Z}_{\ell}^{*}(t_2)$, $\mathbb{Z}(t_2) = \left(\mathbb{Z}_{\ell}^{*}(t_1) \cap \mathbb{Z}_{\ell}^{*}(t_2)\right)$ and $\mathrm{v}_{\ell}(t_1) \not\approx \mathrm{v}_{\ell}(t_2)$. The difference in the information conditions on $\mathbb{Z}_{\ell}^{*}(t_1)$ and $\mathbb{Z}_{\ell}^{*}(t_2)$

goes to update the information stock $\mathbb{Z}^*(t_2)$ such that $\mathbb{Z}^*(t_1) \subset \mathbb{Z}^*(t_2)$ with $v_\ell(t_1)$ destroyed and transformed to $v_\ell(t_2)$ as a new creation for any time structure $t_2 - t_1 > 0 \in \mathbb{T}$. The internal processes are generated by self-excitement, self-correction, and self-learning, under self-organizing internal decision-choice systems to produce continual information. The task now is to design an epistemic process to show how these self-inductions come into being. It has already been stated that the information stock is a property of both matter and energy and the variety is different form of knowing. It must be noticed that the condition $\mathbb{Z}^*(t_1) \subset \mathbb{Z}^*(t_2)$ implies that $\mathbb{Z}_\ell^*(t_1)$ is not destroyed and that $\mathbb{Z}(t_2) = (\mathbb{Z}_\ell^*(t_1) \cap \mathbb{Z}_\ell^*(t_2))$ is an information flow.

5.3.1 *The General Principle of Opposites and Internal Self-dynamics of Varieties*

In the processes of self-excitement, self-correction and self-organizing, self-transforming of varieties, the elements are made possible by the energy process under the condition that matter is always a plenum of forces under tension. In transformation dynamics, every element is described by the characteristics, category, primary category, derived category, polarity, actual-potential polarity, duality, qualitative-quantitative duality, negative-positive duality, negative-positive characteristic sets, internal conflicts, internal force-energy structure, socio-natural decision-choice system, matter-energy self-excitement, paradigms of thought and principle of opposites. The internal conflicts are the result of the competing relation between the internal negative and positive characteristic sets that produce an internal energy and force for the internal dynamic behavior. The manner in which these behaviors are brought into being find expression in the technological space. The relational structures of all these in continuum and unity form the foundations of information production through destruction and creation of varieties. The relational structures of matter, energy and information produce socio-natural technologies for internal transformations and categorial conversions under internal decision-choice systems. The understanding of this behavior requires the study of inter-categorial movements, where such movements are governed by qualitative and quantitative transfer functions. In other words, we must study how an element in one category loses its old categorial characteristics, acquires new characteristics, and migrates from its parent category into a new parent category which is a derived category. The concept of category has been defined, the construct of which is related to fuzzy categorial indicator function in Chap. 3. The current analytical process requires a conceptual definition of categorial conversion in the info-dynamics.

Definition 5.3.1 (*Categorial Conversion*)
Categorial conversion is a process where derived categories emerge from a primary category and where the derived category has a direct or an indirect continuum with the primary category for a reversal process. At the level of ontology,

categorial conversion is a natural process called *ontological categorial conversion*. At the level of epistemology, categorial conversion is an epistemic process called *epistemological categorial conversion*. The distinction between the *primary and derived categories* is established by *categorial differences* as revealed by their qualitative characteristic sets.

As defined, the ontological categorial conversion is the identity as well as the primary category for epistemological categorial conversion which constitutes the derivative in the information-knowledge production process. For any given phenomenon, the epistemological categorial conversion leads to a knowledge discovery and an epistemological info-flow if it is equal to the ontological categorial conversion under epistemic conditionality within the principles of communication and epistemological decision-choice actions [R3.10], [R3.13], [R3.39], [R4.10], [R4.13]. The work in the epistemological space is to discover the varieties of events, states and processes that are taking place in the ontological space where these events, states and processes are independent of the cognitive and non-cognitive elements contained in the ontological space. To illustrate the categorial conversion process, let the elements under the qualitative process in the epistemological space be a set of the form $\{\mathbb{C}_0, \mathbb{C}_1, \mathbb{C}_2, \mathbb{C}_i, \mathbb{C}_\infty\}$ and corresponding to it a time set $\{T_0, T_1, T_2, \ldots T_i \ldots T_\infty\}$. The $\mathbb{C}'$s may be viewed as varieties or categorial varieties, where the time and categories may be written in pairs as $\{(\mathbb{C}_0, T_0), (\mathbb{C}_1, T_1), (\mathbb{C}_2, T_2), \ldots (\mathbb{C}_i, T_i) \ldots (\mathbb{C}_\infty, T_\infty)\}$. The $\mathbb{C}_i'$s may be taken as categorial varieties or individual varieties. Let $\mathbb{C}_0$ be the primary category at the initial time T_0 to be transformed in an infinite time T_∞ where methodological constructionism in a forward time-moving process and in right telescopic into

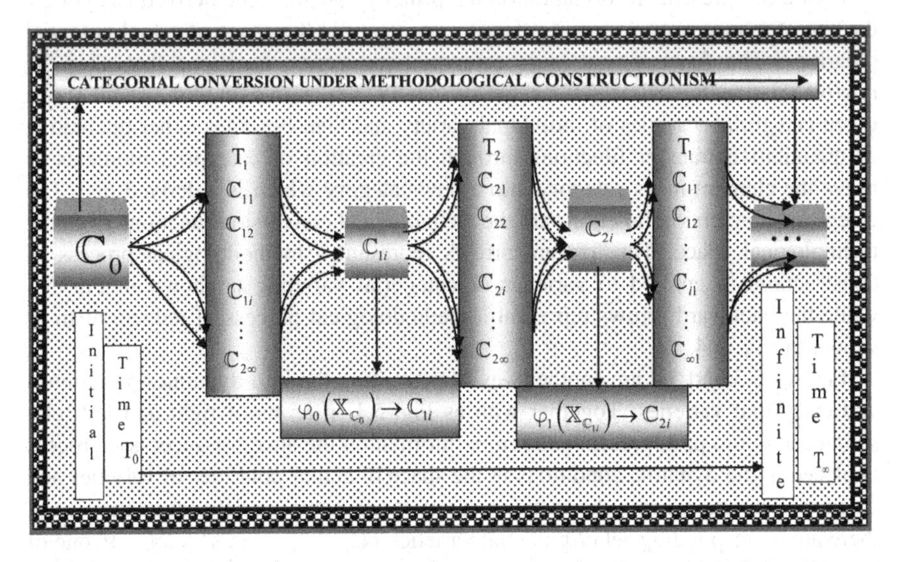

Fig. 5.3 An epistemic geometry of categorial-conversion process from an initial time T_0 to an infinite time T_∞ . That is from parent to offspring

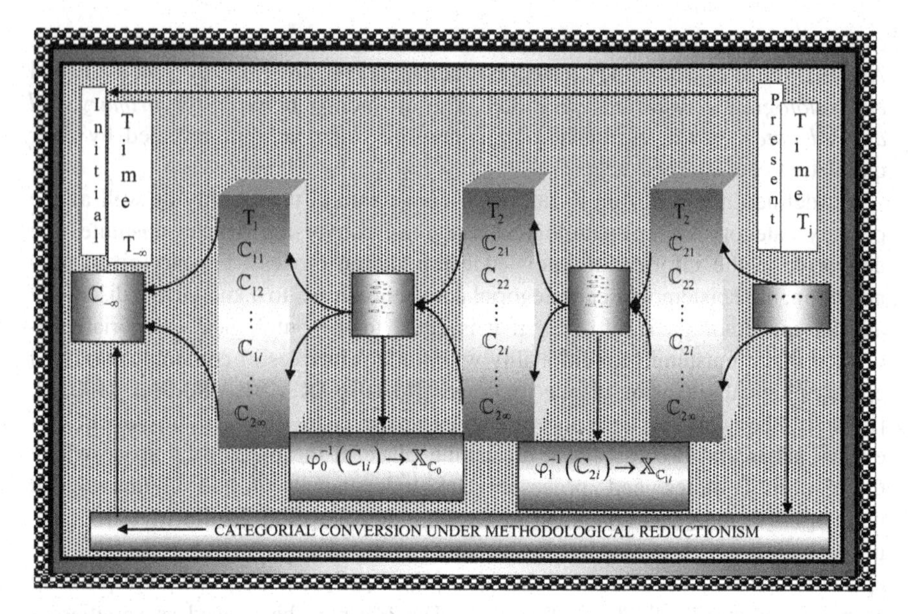

Fig. 5.4 An epistemic geometry of categorial-conversion process from a derived category (offspring) to the primary category (parent) at time T_j to the initial time T_0

infinite time is considered to access the information production structure from an initial T_0 where the initial time may be the present or past. This info-process is presented in Fig. 5.3.

Figure 5.3 presents information on the primary category, the derived categories, the ancestor-successor or parent-offspring processes in the space of actual-potential polarities. The functions $\varphi(\cdot)$ are categorial-conversion functions that alter the relative relationship between the negative set and the positive set of the element in a given categorial varieties of individual variety and then becomes either qualitative or quantitative transfer functions through the creation of categorial moments to move the element in its category into a new category which is a derived one. The categorial-conversion functions are the technological conversion functions from the technological space. The forward arrows represent a methodological process of constructionism applied to varieties or categorial varieties as shown in Fig. 5.3 while the backward arrows represent a methodological process of reductionism as shown in Fig. 5.4.

Figures 5.3 and 5.4 are another way of presenting the info-dynamics of varieties and categorial varieties. Together with as Fig. 5.1 in terms of right and left telescopic view into past, present and future information the ancestor-successor process is placed in the dynamic conditions in the space of actual-polarities of varieties and categorial varieties. From the transformation of any actual variety for example, $\mathbb{C}_0$, there are corresponding set of potential varieties $\{\mathbb{C}_{1t}, \mathbb{C}_{2t}, \ldots, \mathbb{C}_{it}, \ldots, \mathbb{C}_{\infty}\}^{\mathrm{P}}$ one of which may be actualized in the transformation dynamics as shown in Fig. 5.4. It is always useful to keep in mind the relational structure of information stock-flow process and variety-transformation process. Every variety has corresponding identity and every categorial varieties also has corresponding categorial identities. The identities are revealed by the corresponding characteristic dispositions and projected by the corresponding signal dispositions as information through transmissions. Every transformation of a variety or categorial variety presents a change in the information stock-flow conditions which show some regularities in the universal existence in the information space. These regularities may be viewed in relation to time and info-conditions of the varieties and categorial varieties which are associated with the dynamics behavior of information stock-flow conditions in the universal existence of varieties and categorial varieties.

5.3.2 Some Essential Postulates of Information, Time and Events in Ontological Space

The analytical point that is being presented here is that ontological characteristics are generated by matter in various forms to inform the presence of different ontological objects. Matter generates qualities which establish distributions of *categorial varieties*. The ontological characteristics constitute the primary category of information from which other categories of information can be shown to be derivatives by examining the interplay of characteristics, signals, quality and quantity of ontological objects. Any definition of information must involve characteristics and signals that present qualitative and quantitative varieties of elemental existence. It is useful to keep in mind that behind every signal disposition, there is a characteristic disposition of matter or its surrogate which is traceable to matter by methodological reductionism. There is no signal or information without a characteristic set. It is the characteristics that allow cognitive agents to speak of events and outcomes, and to be informed and acknowledge that an event has occurred. It is a change in the characteristic-signal disposition over time that informs cognitive agents that a transformation has occurred. It is here that the problem of understanding the relational continuum and unity of matter, energy, information and time presents some important cognitive challenges in the epistemological space. It is also here that the study and research on *info-dynamics* become increasingly important as a subject area science and philosophy needed for the advancement of frontiers of different knowledge areas. The subject area of *info-dynamics,* its possible contents and usefulness has been discussed s and explained in the previous chapters of this monograph. A reflection here is useful to understand the fundamental laws of info-dynamics.

A question may be asked as to how one knows that an event in the traditional sense of probability has occurred over the epistemological space. This question holds irrespective of whether one subscribe to objective or subjective concept of probability. It also holds whether one is working with the classical paradigm of thought or fuzzy paradigm of thought. If one has a coin of the same faces, the flipping of the coin will present no distinguishable outcome no matter the number of times the coin is thrown. The reason is that the faces present no variety and hence no information from the outcome. By designing the coin with two different faces, the principle of opposites is established to create dualism with excluded middle and hence the faces have been assigned differential characteristics such as head or tail that may be related to conditions of potential, possibility, probability and actual in terms of characteristics and signals that come to be generally branded as information which links matter and energy. At the level of ontology, this information is complete and exactly transmitted such that there are no uncertainties, accidents and risk but a continual ontological categorial conversion where different manifestations of matter are brought into existence with a creation of different manifestations of energy and continual accumulation of information with info-flow.

Definition 5.3.2.1 (*Info-dynamics*)

Info-dynamics is an area of study concerned with the study of the dynamics of categorial varieties with relational continuum and unity of matter and energy in the spaces of ontology and epistemology. It is also concerned with the study of statics and dynamics of ontological-epistemological characteristics, characteristic clustering, variety identification, category formation and transformations in qualitative and quantitative dispositions with neutrality of time. It works with both macro-varieties and micro-varieties of matter and energy in terms of identification, distinction and similarity. Info-dynamics apply to all areas of knowledge production under both static and dynamic conditions in the study of categorial variety and categorial conversion. The essential contribution of info-dynamics is its universality to all areas of informing, knowing and learning, where identification, distinction and similarity present themselves in varieties and categories. It is intended to study the common philosophical and mathematical properties of inputs into information-knowledge production in all actual and potential fields. It involves the development of concepts, measurements and properties of information and time that will present a relational continuum and unity of socio-natural transformations of varieties and categorial varieties for the understanding of information-knowledge-production system. Comparatively the info-statics study the definition of information and its relationships to, properties of matter-energy varieties in a fixed time. This initializes the study of info-dynamics regarding the transformations of varieties and categorial varieties.

5.3.2.1 Laws and Postulates of Info-dynamics

Given the framework of info-dynamics, a number of important analytical postulates are derivable. These postulates will be stated and discussed.

Postulate 5.3.2.1: The Law of Information Indestructibility
Once an information is created, it cannot be destroyed or altered at the level of ontology where such information is reflected by characteristic-signal disposition.

Postulate 5.3.2.2: The Law of Continual Information Creation and Accumulation
Information is continually being created and accumulated at the dynamics of categorial variety and ontological transformation in the natural decision-choice processes.

Postulate 5.3.2.3: The Law of Information Objectivity
At the level of ontology, information is complete and exact such that there are no uncertainties, accidents and risks in the process of continual creation of categorial varieties.

Postulate 5.3.2.4: The Law of Continual Information Disequilibrium
Information is always at disequilibrium state where there is no stock equilibrium or a flow equilibrium at the levels of inter-relations, intra-relations of ontological objects and info-dynamics.

Postulate 5.3.2.5: The Law of Cost-Benefit Process
Every pre-successor (ancestor or parent) is a real cost and every successor (off-spring) is a real benefit in the info-dynamics of varieties and categorial varieties where the ancestor is transformed to give birth to the offspring or the successor.

5.3.3 Comments and Reflections on the Laws and Postulates

The law of information indestructibility is the *first law of info-dynamics* and points to the conditions that once a categorial variety emerges out of ontological dynamics its identity is sealed by its characteristic-signal disposition and that a categorial conversion of any ontological object, while creating new characteristic-signal disposition is incapable of destroying the characteristic-signal disposition that formed the information basis of the identity of the ontological object of the previous category. Information about any categorial variety is formed and stored and cannot be destroyed. The first law of info-dynamics is the *law of permanency of ontological identities*" in histories. It is also the law of preservation of characteristic-signal disposition.

The law of continual creation and accumulation of information is the *second law of info-dynamics* and constitutes the ontological history as a universal property of matter and energy in the past, present and future. The second law of info-dynamics is the law of *permanency of categorial conversion* or dynamics of categorial variety. It affirms the existence of varieties of matter and energy and continual differentiations through categorial conversions in relational continuum and unity in quality-quantity duality within the actual-potential polarity. The *theory of categorial conversion* is about the study of the necessary conditions of internal self-transformation and differentiation of matter through the interplay of matter, energy and information [R17.15]. The sufficient conditions for internal self-transformation and differentiation of matter through the interplay of matter, energy and information is the study of *the theory of Philosophical Consciencism* [R17.16].

The law of information objectivity is the *third law of info-dynamics* which ensures noiseless transmission of information for categorial conversion in the ontological space among ontological object (from IT to IT). The law of information objectivity is an affirmation of ontological identity as well as the primary category of awareness through acquaintance, knowing and knowledge formation over the epistemological space. At the level of ontology, as it is being presented, there is no distinction between information and knowledge as inputs into the *ontological decision-choice processes*. Information is knowledge and knowledge is information both of which are represented and transmitted as characteristic-signal dispositions among ontological objects in ontological space. Here, it is the qualitative characteristics of universal objects that establish ontological differences and similarities at inter-relational levels, where the collection of objects of qualitative sameness constitutes a qualitative categorial variety and hence establish a distribution of qualitative inter-categorial differences.

These *inter-categorial differences* present a structural distribution of qualitative ontological varieties in the universal system. It is these categorial differences that allow one to know the moon as different from the sun and to know a cow as different from a person and energy as a derivative of matter. The common defining factors among all these ontological objects are *characteristics* and *signals* which reveal the nature of the characteristics as surrogate representation of matter in different forms. Here, there are interactive processes of quality, quantity and time such that within every qualitative categorial variety there is a distribution of quantitative categorial varieties at the level of ontology. This is the intra-categorial differences. For example, a collection of trees constitutes a qualitative categorial variety in which there is a distribution of quantitative varieties defined by linguistic numbers such as small, medium and big that may be reduced to arithmetic numbers. Alternatively viewed, the universe of things is composed of distributions of qualitative categories in relation to the properties of qualitative dispositions, and quantitative categories in relation to the properties of quantitative dispositions. It is also here, that one encounters the principle of opposites in terms of collections of universal dualities with relational continuum and unity in the universal actual-potential polarities.

The law of information disequilibrium is the *fourth law of info-dynamics* which ensures the notion that the only thing that is permanent in the ontological space is continual transformation through categorial conversion of categorial varieties, where the categorial conversion is continually manufacturing different forms of characteristic-signal disposition and categorial varieties. The continual creation of information simply means that the internal self-transformation of matter creates new varieties of matter, and these new varieties of matter gives rise to new varieties of sources of energy from the same matter and energy, and corresponding to them arise new characteristic and signal dispositions of varieties and categorial varieties with changing varieties that present information flow to update the stock of information in continual disequilibrium state. Matter and energy are never destroyed but continually being transformed to create new information over the spaces of ontology and epistemology. The law of cost-benefit process is the fifth law of info-dynamics which ensures the notion of input-output process in transformations where the inputs are pre-successors which are the real opportunity costs forgone to bring new real benefits from the successors. This real cost-benefit process also ensures the continual variety transformations with indestructibility of matter and energy in universal existence where matter in collaboration with energy from within under principle of information transforms itself into varieties without any loss. The real cost-benefit process is always in relational continuum and unity in terms of real opportunity costs where real costs are benefits and real benefits are real costs in the socio-natural transformation dynamics. It is important to relate the real cost-benefit processes to the fundamental laws in thermodynamic, electrodynamic, electromagnetism and many areas of scientific theories and practices where matter is both input and output in its internal dynamics for the info-dynamic system. Nothing is knowable over the epistemological space without information and without information there exist no varieties and alternatives, and hence the practice of decision-choice actions are impossible over the epistemological space. Similarly, without varieties and categorial varieties language development is impossible.

5.3.4 Stocks, Flows and Information Disequilibria

It is here that some important distinctions arise between information on one hand, and matter and energy on the other hand; and between info-dynamics and thermodynamics, as well as other relevant laws of physics, chemistry and biology. In general epistemic analytics, there are stocks and flows in the analysis of statics and dynamics of states and processes. There are stocks and flows of matter, energy and information as common behavioral factors on the basis of which the concepts of equilibrium, disequilibrium and stability are created for analytical understanding of behaviors of elements in the universal object set. There are the concepts of *stock equilibrium* and *flow equilibrium*. While there are stock and flow equilibria in matter, energy and thermodynamics, electrodynamics, economic dynamics, bio-dynamics and other forms of dynamics, there are no stock and flow equilibria in information and info-dynamics. This condition of lack of equilibrium in information and its dynamic behavior places information as a special property of matter. The lack of equilibrium in information is directly transmitted to knowledge behavior over the epistemological space.

To understand this powerful analytical differences among information behavior and behavior of matter and energy and their derivatives, it is useful to pose some questions and answers. What are the concepts of stock, flow and equilibrium, and how do they relate to matter, energy and information, and how do they help in understanding the general concepts of statics and dynamics in the information-knowledge-production systems. It is useful to keep in mind in the general epistemic analytics that there are, *organic variables* and within each organic variable are *macro-variables* and *micro-variables*. Matter, energy and information are organic variables whose static and dynamic behavior are of interest in relation to the concept of equilibrium. The categories and varieties are cast in macro-variables and micro-variables, the static and dynamic behaviors of which are of interest in relation to categorial conversion and transformations of varieties and categorial varieties within the organic variables as internally induced.

Definition 5.3.3.1 (*Stock and Flow variables in Relation to Matter, Energy and Information*)

A variable is said to be a *stock* if its quantity is measured in a time point. It has no time dimension, however, it has time reference points for epistemic assessment, measurement and computation. A variable is said to be a *flow* if its quantity is measured only in terms of periods of occurrence. It has time dimension, as well as time reference points for all epistemic assessments, measurements and computations.

Matter, energy and information are organic variables; and as organic variables have stocks but not necessary flows since a flow variables have magnitudes in interval of time while stock variables have magnitudes over points of time. In this respect, care must be taken in algebraic specifications and operations of stock-flow variables in all analytical works in both exact and inexact knowledge spaces.

Definition 5.3.3.2 (*Equilibrium*)

An organic variable is said to be in a *flow equilibrium* when the net flows are zero. It is an equilibrium defined for a time dimension and as a maintenance of stock equilibrium, given the flow equilibrium conditions. An organic variable is said to be in a *stock equilibrium* if the net additions to relevant stock variable are zero. The stock equilibrium, therefore, is an equilibrium at a point with no flow additions.

5.3.4.1 Transformations, Stocks, Flows and Information Disequilibria

In general, flow equilibrium describes conditions of short run such as temporary equilibrium state. The stock equilibrium backed by flow equilibrium describes conditions of long-run equilibrium. The following epistemic carefulness must be taken. The first is the distinction among organic variables, macro-variables and micro-variable and the role they play in the understanding the relational structure and unity among matter, energy and information. The macro-variables may be seen as categories and micro-variables may be seen as individual variety within the relevant organic variable. Given these variables, it is useful to note that a static state of an organic variable in categorial conversion does not imply a static state in categorial conversion for either the macro-variables, micro-variables or both. In other words if, for example, the total matter is in categorial conversion equilibrium, it does not imply that the new categories and varieties are not been created to generate new sources of energy and information. The variety transformation process is such that one type of variety of matter or energy is simply converted to another type of variety of matter or energy without a loss. Similarly, a transformational dynamic state for any of the individual macro-variables and micro-variables does not imply a transformational dynamic state of a relevant organic variable. In matter and energy, the organic variables are always in stock equilibria while the number of varieties is in disequilibrium. A stock equilibrium for an organic variable does not imply a stock equilibrium for either the macro-variables or micro-variables and a stock disequilibrium for an individual macro-variables and micro-variables does not imply a stock disequilibrium for any relevant organic variable. In other words, the universe may be closed in the sense of being in stock and flow equilibrium and hence not expanding, this does not mean that other celestial bodies and solar systems are not been created by categorial conversion within the universe where new varieties and categorial varieties are formed in the sense that the universe is infinitely closed under continual transformation with the number of varieties and categorial varieties continually changing. In fact this is the nature of population dynamics and ecological dynamics.

The collection of these analytical ideas is important in understanding the definition of information as characteristic-signal disposition and as a property of matter, as well as categorial varieties driven at the level of ontology. It is also important in understanding knowledge-creation as a self-correction system in relational continuum and unity of informing, knowing, teaching and learning when one links the ontological space to epistemological space. The law of information disequilibrium as the *fourth law of info-dynamics* is extremely powerful in that the creation and the *flow of information* is always in a positive direction where the stock of the volume of information is continuously increasing by its flow through the creation and recreation of varieties and categories under continual destruction of old varieties and with continual formation of new characteristic-signal dispositions which are indestructible within the organic and specific transformation-substitution duality.

The stock-flow conditions of info-dynamics connect the time-trinity of past, present and future as philosophically represented by an African symbol of *sankofa-Anoma* which illustrates the continual connectivity of the past, present and the future in creation and destruction of varieties and categorial varieties with their transformations and information generations. The info-dynamics presents itself in dynamics of *intra-categorial* and *inter-categorial* varieties that helps to explain continual transformations through construction-destruction process as the combined works of matter, energy and information. The info-dynamics, within the *sankofa-Anoma* tradition in transformation and information-generation processes, is revealed in the analytical works of forecasting from the past, prediction from the present into the future, discounting from the future into the present and surprise in info-dynamics that generates information input-output structure for current decision-choice systems. It is useful to keep in mind that the past is known to cognitive agents within their present context by its *stock of information,* while the future is revealed to cognitive agents within their present context by its *possible flow of information* that the methods and techniques of cognitive agents can revealed to them.

The stocks of information (info-stocks) relative to the represent the identities of the past varieties and categorial varieties including those of the present whether known or not over the epistemological space, while the flows of information represent the future varieties and categorial varieties that the transformation processes will engender, all in reference to the present state within the past-present-future connectivity. This is the inner essence of the universal information-production machine of transformation-substitution continuity of varieties and categorial varieties that maintains the organic matter-energy stock-flow equilibria with specific and universal information stock-flow disequilibria. The transformation-substitution processes of varieties and categorial varieties are the conceptual keys to explain the conditions of universal matter-energy stock-flow equilibria and the universal changing conditions in the never-ending socio-natural creations. It is this nature of info-dynamics that gives meaning to the idea that the future resides in the present, while the present resides in the past which is connected to the future.

The combined stock-flow information is the input into decision-choice systems of dynamics of socio-natural systems of actual-potential polarities. It is also a framework to understand that the whole of the universal system is a collection of socio-natural problem-solution processes where every solution creates its own new problem requiring new solution to bring new problem. It is this collection of problem-solution processes that defines the dynamics of the collection of socio-natural actual-potential polarities in universal existence and gives meaning to human life. These are the epistemic foundations of info-dynamics and in collaboration with info-statics provide the structure of the general information theory that deals with the qualitative and quantitative defining structure of the concept of information to fix its phenomenon, the requirements of its measurement and its self-creative force at the level of contents of varieties and categorial varieties.

5.3.4.2 On the Theory of Info-dynamics, Categorial Conversion and Philosophical Consciencism

Info-dynamics may be viewed as the branch of categorial conversion and Philosophical Consciencism concerned with information-knowledge production systems and their relationships to socio-natural decision-choice systems that bring about socio-natural destruction-construction actions of varieties and categorial varieties under transformation-substitution principles with or without intentionalities in the decision-choice system. The aspects of categorial conversion provide the necessary conditions that project necessities defined in the possibility space of the variety transformations. The aspects of Philosophical Consciencism relate to the decision-choice systems that provide the sufficient conditions which project freedom defined in the probability space of the variety transformations. The info-dynamics defines both microscopic information processes such as intra-categorial conversions of varieties and macroscopic information processes such as inter-categorial conversions of categories. It is concerned with the dynamics of information flows and their impact on the forces of decision-choice actions on varieties and categories over both the ontological and epistemological spaces given the info-static structure.

Info-dynamics applies to all areas of socio-natural activities without an exception. The subject matter of info-dynamics relates to matter and energy for the production of intra-categorial and inter-categorial varieties and categorial varieties. The qualitative and quantitative directions of the intra-categorial variables and inter-categorial are only constraint by their environments that define the necessary conditions for transformation. This necessary conditions are provided by the theory of info-statics which specifies the initial conditions for change. The sufficient conditions implied by the info-dynamics find expression in decision-choice activities. It is different from thermodynamics which has deals with equilibrium and describes the bulk behavior of the body and not the microscopic behavior of the characteristics. It is useful to keep in mind that the concepts of equilibrium and disequilibrium acquire differential meanings in the theory info-dynamics and the theories of thermodynamics, electrodynamics, economics bio-dynamics and many others. The stock-flow variables in the info-dynamics must be relate to the dynamics of the system of actual-potential polarities due to socio-natural decision-choice activities. The info-dynamics deals with both the micro and macro variable separately and jointly because of relational continuum and unity. In the development of the theory of info-dynamics, one must examine in separation and unity the decision-choice activities over the ontological and epistemological spaces. It is on the basis of info-dynamics that the information related concepts are examined over the epistemological space.

Chapter 6
Info-dynamics and Some Knowledge Areas

The theory of info-dynamics represents a unified theories of dynamics as such, it is useful now to examine the relational structure of the theory of info-dynamics to other dynamical systems over the epistemological space. The theory of info-statics involves three organic dimensions of matter, energy, information which together constitutes the description of the initial state of being for epistemic analysis. It, thus, involves the three dimensions in addition to a fourth dimension which is time. There are three important categories for comparative study and analysis. The organic categories are *matter-time category*, *energy-time category* and *information-time category* in a relational continuum and unity in the sense that energy and information are embedded in matter, where the identities of matter and energy are revealed by information as a category. Another way of saying the same thing is that matter and energy are indistinguishable without information. These three elements constituting the organic category of dynamical systems are applicable to all areas of existence. Matter-time dynamics and energy-time dynamics are embedded in information-time dynamics the explanatory and prescriptive behaviors of which find meanings and expressions in the principle of opposites, where the opposites are composed of systems of actual-potential polarities and negative-positive dualities with relational continua and unity. This may seem absurd but it is simply the nature of the universal process. The categories of matter-time dynamics and energy-time dynamics are interpreted within the category of information-time dynamics. Nothing is knowable, teachable, learnable, transmittable, selectable, decidable or communicable over the epistemological space without information. Life has no existence in an information-less environment.

Let us keep in mind that information is multidimensional and multilateral relations among elements in the universal object set Ω composed of things, objects, states and processes where a generic element is $\omega \in \Omega$. The mutual relations find expressions in the source-destination dualities in terms of transmissions and communications of the characteristic dispositions in expressive continua and unity. The nature of these relations depends on the quality and quantity of the contents transmitted and communicated and the interpretations of the contents through the

K.K. Dompere, *The Theory of Info-Dynamics: Rational Foundations of Information-Knowledge Dynamics*, Studies in Systems, Decision and Control 114, https://doi.org/10.1007/978-3-319-63853-9_6

acquaintances with the signal dispositions. The quality relates to the degrees of vagueness of the contents and the quantity relates to the limitations of the degrees of fullness or completeness of the volume of the contents as received and conceptualized by the destination from the source. The degree of the vagueness is measured in the *possibility space* in terms of subjective phenomenon in the clarification of the vagueness into degrees of exactness. The degree of completeness is measured in the *probability space* in terms objective phenomenon in clarification of the degree of completeness of the volume of the content received.

The relational types of information among elements in either bilateral or multilateral mode will be affected by the degrees of vagueness and completeness of the contents transmitted or communicated in the source-destination process between the sets of elements of ontological and epistemological spaces and among the elements within the epistemological space. The information as relations among things is universal without exception. Even in a single object world, the object is in relation with itself induced by an information process. It is this information among things that establishes relations among things and create problem-solution processes changing relation and give meaning to life and existence.

The logic of the general information definition (GID) is such that all things are defined and related by characteristic-signal dispositions that establish their identities with the corresponding varieties and their mutual behaviors. This logic of the variety is universal that defends, as well as preserves, all arguments over all things as constituting differential sets of characteristic dispositions, the presence of which are made known by signal dispositions to establish multilateral or bilateral relations, and that the phenomenon of information is affected by this scope of logic relating to identity and variety of information types. This logical system of identity-variety-relational structures leads to the solution to the *identification problems* including types of elemental identities and types of relations in general. The identification problem is general to knowing learning and deciding. It is not limited to science and scientific know-how. It involves the nature of existence of life-death duality, destruction-construction duality under the principle of opposites. The concept of types of relations spanned by information is extremely very important to understand statics and dynamics of things in quantity-quality-time multidimensional time phenomena. It is on the basis of the concept of information relations that the *concept of entropy* finds its measurable values and meanings. Let us keep in mind that there are infinite modes of relations between and among the objects in $\omega \in \Omega$ and phenomena $\phi \in \Phi$. The set of infinite modes of relations generate a set of entropy values. It is, therefore, useful to examine the relationship of info-dynamics to thermodynamics, electrodynamics, energetics, electromagnetism, linguistic dynamics, and entropy and information relations.

6.1 Info-dynamic Reflections on Thermodynamics

One may recall from the previous chapters that the universal existence has three pillars of matter, energy and information in an inseparable unit. This unified unit is shown as an epistemic model representation in Fig. 6.1. The existence of matter and energy implies the existence of information, the definition of which must be clearly established in generality to indicate the existence of differential forms of matter and energy in varieties with defined identities. In most simple form, thermodynamics belongs to physics as one of its many branches. These branches of physics are partitioned and revealed by their characteristic-signal dispositions into varieties with defined fuzzy identities. The thermodynamics may also be partitioned into another set of characteristic-signal dispositions into varieties with defined fuzzy identities. The basic form is that it is devoted to the study of heat and temperature and how heat and temperature relate to the production of energy, entropy and pressure, and how they are mutually related and partially provide a descriptive notion of a body of matter with quantitative and qualitative motions. The thermodynamics relates to the behavior of varieties of energy and how these varieties are derivatives from matter where matter constitutes the primary category of existence. There are general constraints that affect the relational structure of

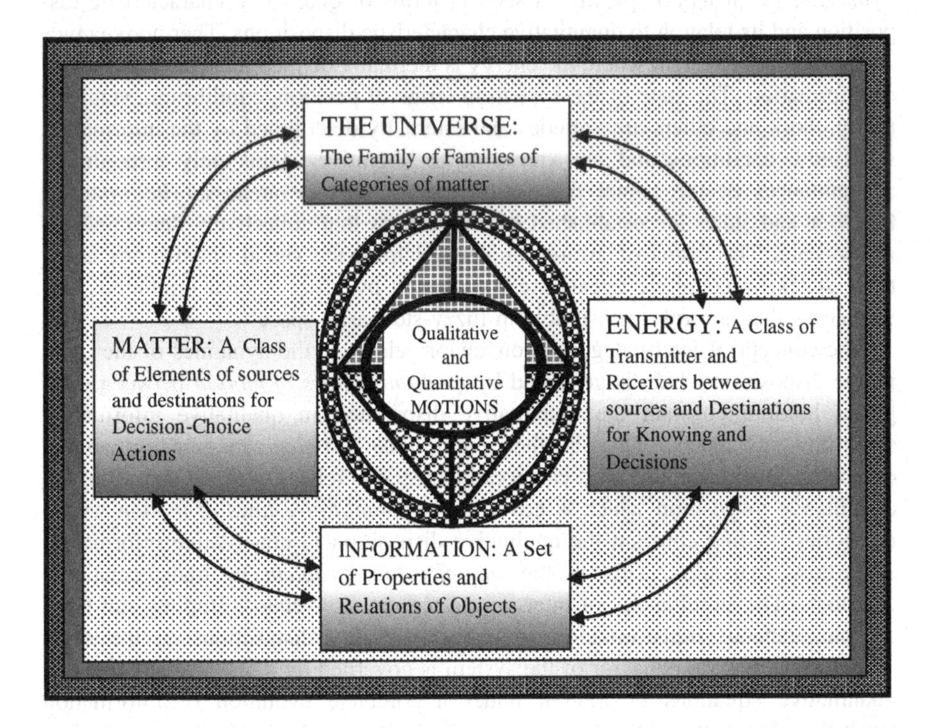

Fig. 6.1 Epistemic geometry of a relational structure of matter, energy information and transformations of varieties

thermodynamic energy to matter. The constraints find meanings in information system and are expressed over the epistemological space as (four) laws of thermodynamics, where the focus is on the macro-relations to heat generation to energy production and to force for internal and external qualitative and quantitative motions. Every ontological or epistemological relation is an information relation that finds meaning in matter and generates action and reaction with decision-choice responses with either intentionality or non-intentionality. When the action of knowing arise, the information dimension of the universal existence constitutes the primary category of knowledge about matter and energy even though matter is the primary category for variety transformations.

In the general knowledge-production process, thermodynamics is about the study and uses of knowledge on thermal properties of matter to bring about inter-categorial and intra-categorial conversions of varieties. The study includes thermal energy transfer through all kinds of media, where one can speak of biological thermodynamics, biochemical thermodynamics and many others. The central focus of the framework of thermodynamics relative to information is a system and its definition leading to the development of scientifically conceptual and analytical areas of socio-natural systemicity and complexity. Fundamental to all systems and complexities are states, energy processes, control-energy operations, the boundaries of the system and equations of motion governing the system's dynamics, given its complexity as seen in terms of qualitative characteristic disposition and its relation to quantitative characteristic dispositions. Thermodynamics is part of energy system where the energy is thermally defined relative to the nature of information. The state of the system in terms of position, equilibrium, stability, utility, work and others are varieties and internally determined by the information stock-flow conditions and are knowable to cognitive agents only through the acquaintances of the signal dispositions to establish awareness. The system's behavior and states are established by microstates that collectively determine the macro-state. The forms of awareness derived from the signal dispositions about these states by cognitive agents are constrained by observational limitations and vagueness that are truly measurable in fuzzy-stochastic space.

The concept of inter-categorial conversion relates to the dynamics of the qualitative disposition which is governed by *qualitative laws of motion* between categories holding constant the quantitative disposition in qualitative information stock-flow dynamics. This is the *quality-time problem* in transformational dynamics. The concept of intra-categorial conversion relates to the dynamics of the quantitative disposition which is governed by *quantitative laws of motion* within categories holding constant the qualitative disposition in quantitative information stock-flow dynamics. This is the *quantity-time problem* in transformational dynamics. The simultaneity of inter-intra categorial conversions implies that there are simultaneous transformations of quantity-quality dispositions of the same variety such that the behavior of the system is governed by a set of qualitative and quantitative equations of motion under a complete evolution of information stock-flow dynamics. In this way, one must simultaneously deal with quantity-time problem as well as a quality-time problem, where the particle under space time motion is also under internal transformation of its qualitative variety into different and unknown variety. This is the *quality-quantity-time problem* in transformational dynamics.

Every qualitative or quantitative state of an element is a variety defined by a characteristic-signal disposition, where the characteristic disposition establishes the identity of the state of the system and the signal disposition establishes the possible awareness of the states. In terms of info-dynamics, there are two way of viewing the behavior of any system in terms of either micro-conditions or macro-conditions or both. These micro-macro conditions may present themselves in terms of either qualitative disposition or quantitative disposition or both. Any previous state is defined by its information stock, while the inter-states are defined by info-flow. The path of the dynamics of any system is a time-ordered set of varieties that constitutes an enveloping of information stock-flow as the info-dynamic system. Each variety under transformation dynamics is either qualitative-time disposition holding the quantitative disposition constant or quantitative-time disposition holding qualitative disposition constant or qualitative-quantitative-time disposition.

The info-dynamic system defines all possible relations among states and more since the info-dynamics has no equilibrium. The equilibrium or non-equilibrium of any system is information stock-flow production in terms of relational structures. There is no uncertainty in this process of the systems state. The problem of uncertainty in all relational dimension emerges over the epistemological space in which cognitive agents operate through acquaintance to establish an awareness. In other words, the systems dynamics in both qualitative and quantitative dispositions are information processes in stock-flow conditions that may be vaguely and incompletely known to generate the dynamics of *information deficiency*, composed of *stochastic-information deficiency* and *fuzzy-information deficiency* and fuzzy-stochastic information deficiency. It is in this respect that the state of the system specified as equilibrium, stability, internal self-destruction, self-correction, self-excitement, self-correction and others are either random, fuzzy, fuzzy-random or random-fuzzy variables and related to system's entropy viewed in terms of information-knowledge processes which are defined over the space of epistemological dynamics and not over the space of ontological dynamics.

6.2 Info-dynamic Reflections on Electrodynamics

In a simplest form, electrodynamics is devoted to the associated charged particles in motion and changing electric and magnetic fields, and how the dynamics of a charged particle and magnetic particle affect an inter-categorial and intra-categorial motions in production of varieties and hence production of info-flows. The inter-categorial and intra-categorial motions are such that the electromagnetic field is related to everything and with varying degrees of force and impact. The varying degrees of force and impact are related to the internal and external transformation of varieties in such a way that the electric and magnetic fields are constantly changing within each variety. The complexity of electromagnetics is such that categories of its subject matter are many and continually changing as the field of matter alters in varieties. All these forms of electromagnetic system relate to the behavior of varieties of energy and how these varieties are derivative from matter which constitutes the primary category of energy. These energy varieties are defined by characteristic-disposition components (contents) of information to create identities while the signal disposition component (transformation) of information creates awareness of the identities to the ontological objects.

The relational structure of electrodynamic varieties, just like thermodynamic varieties to matter is established by a set of general constraints. The set of these constraints also finds meaning in the information system and expressed in the laws of energy relevant to the internal and external equations of motion in the dynamic space of qualitative-quantitative dispositions in reference to the time dimension and in relation to the distribution and shifting distribution of the characteristic-signal dispositions. In terms of the theoretical framework that is sought in the general knowledge-production process, electrodynamics is about the study and utilization of electromagnetic properties of matter to effect inter-categorial and intra-categorial conversions of elements to create forms of varieties with corresponding identities as revealed by characteristic-signal dispositions.

The study of electrodynamics, like that of thermodynamics, includes convertibility of energy into different forms and the corresponding works, and as well as the transferability through all kinds of media. In this respect, one can meaningfully speak of biological electrodynamics as well as bio-economic electrodynamics, institutional electrodynamics and other forms since all existing forms are derivatives of matter and all changes and transformations are induced by energy. The control focus of the electromagnetics is its production of force and work which bring about qualitative and quantitative laws of motion to continually produce internal transformations of varieties and categorial varieties generating information stock-flow dynamics from the past through the present and to the never-ending future, the realization of which is uncertain and unknown to cognitive agents over the epistemological space. This future uncertainty resides in every variety and is related to specific entropy within the general entropy of universal existence that finds primacy in matter as conceptualized over the epistemological space. Just as nothing is knowable without information, nothing is changeable without energy and no variety exists without matter, energy and information. Every conceptual characterization of uncertainty is future defined and every future resides in uncertainty over the epistemological space.

6.3 Info-dynamic Reflections on Energetics

From the concepts of thermodynamics and electrodynamics, an attention is turned to the general concept and subject area of energetics. Here, energetics is seen in terms of a relational structure of all forms of energy and its static and dynamic behaviors in matter, broadly defined. Energetics, therefore, is in some general sense the study of forms of energy to bring about inter-categorial and intra-categorial conversions of varieties of various forms of matter, energy and information. All forms of existence are varieties with identities and all categorial conversions of forms of existence are transformations in the quantity-quality space with neutrality of time as a fourth dimension to matter, energy and information. Without energy transformation is impossible and without matter, the existence of energy is undefinable, and without matter information is non-existence while without energy-matter interactions information flows, in terms of variety destructive-creative process, is non-conceivable. The interactive processes of matter, energy and information destroy exiting varieties and create new varieties from within under the active work of energy by changing the distribution of characteristic-signal dispositions, the results of which update the stock of information through the flow of information. The work of energy must be seen in terms of varieties of energy as represented in Fig. 6.2.

The theory of energetics has a continuity with the theory of transformation which may be divided into sub-theories of *Categorial Conversion* and *Philosophical Consciencism*. The theory of energetics may also be divided into sub-theories of energy-statics (*energestatics*) and energy-dynamics (*energedynamics*). The theory of categorial conversion establishes the necessary conditions of transformations of varieties and categorial varieties in terms of inter-categorial and intra-categorial convertibility of socio-natural varieties. The theory of Philosophical Consciencism establishes the sufficient condition of transformation of varieties and categorial varieties in terms of inter-categorial and intra-categorial convertibility of socio-natural varieties. The theory of *energestatics* may be related to the conditions of potential energy and hence defines the necessary conditions of power generation and work in systems. The theory of *energedynamics* establishes the sufficient conditions to bring about power and work in systems. Energestatics relates to categorial conversion and then to info-statics of varieties and categorial varieties at any point of time. The static conditions of varieties establish info-statics and *energe-statics* providing the needed conditions for variety identification in terms of identity, difference and similarity Energedynamics relates to Philosophical Consciencism and then to info-dynamics over infinite time domain. As presented, the subject matter of energetics cover a broad disciplines involving necessary and sufficient conditions of statics, dynamics and stability of elemental varieties at various levels of work to destroy existing varieties, create new varieties and modify existing varieties.

From the point of view of the epistemic frame that is being sought here, energetics has a general relationship to information through the interactions of energy and matter at various forms and linked to information stock-flow dynamics over infinite time. Here, energetics is not simply related to property of matter but how it is related to conditions of internal destruction of existing forms of varieties and the creation of new forms of varieties for continual production and reproduction of information, where information is defined as a characteristic-signal disposition and the distribution of socio-natural varieties has a corresponding distribution of characteristic-signal dispositions where energy is continually active to produce info-dynamic relations among varieties for new forms. It may be noted to note that both the sets of socio-natural ontological varieties, categorial varieties and characteristic-signal dispositions are infinite and increasing with a continual expansion of the space of complexity. It is in this respect, that the epistemological space must be viewed as a small sub-space of the ontological space. Technically, therefore, energy stock-flow dynamics is related to the effects of destruction of existing forms of socio-natural varieties without the destruction of the corresponding characteristic-signal dispositions, as well as related to the creation of new forms of characteristic-signal disposition to increase the information stock conditions.

The energy stock-flow dynamics is directly related to information stock-flow dynamics, where energy like matter cannot be created or destroyed. Information, however, cannot be destroyed but is continuously being created through the matter-energy interactive behavior within the space of differentiations and varieties over all infinite past-present-future times. The identification of forms of varieties and categorial varieties of matter and energy as well as forms of information belongs to the subject matter of info-statics under *energestatics*. The identification of the process of destruction of existing forms of varieties and the creation of new forms of varieties belongs to the subject matter of info-dynamics under *energedynamics*. The subject matter of info-statics is taken up in the monograph devoted to its theory [R17.17]. The current monograph is devoted in dealing with the theory of info-dynamics and its defining epistemic framework for the understanding of conditions of knowledge-decision dynamics to effect variety transformations.

Energy as a property of matter presents itself in different forms in varieties and categorial varieties. These forms of energy relate to different forms of power and force that do different forms of work destroying varieties, modifying varieties and creating varieties. This process is the foundational behavior of info-dynamics. At the current time, there are about twelve forms of energy that may be identified as shown in Fig. 6.2. The identification of these forms and new forms of energy is also part of the general identification problem in knowledge production, the solution of which finds expressions in information that brings into focus their characteristic and signal dispositions. These energy forms are intimately interactive and produce qualitative and quantitative motions for elemental transformations of socio-natural varieties. All forms of energy play major roles not only in affecting the internal and external structures of properties of all ontological objects and objects of social forms but transform them qualitatively and quantitatively, as well as affecting their relational structures under conditions of unity. In other words, all universal objects exist in relational continuum and unity in varying degrees of interdependency under the general principles of opposite.

When it is said that matter exists under plenum of forces in tension, this statement is in a direct appeal to the relational structures of forms of energy behavior and their effects as are currently known and supported by our current scientific knowledge, where matter takes its forms as important intermolecular forces among the individual molecules in matter itself creating relational interactive continuum

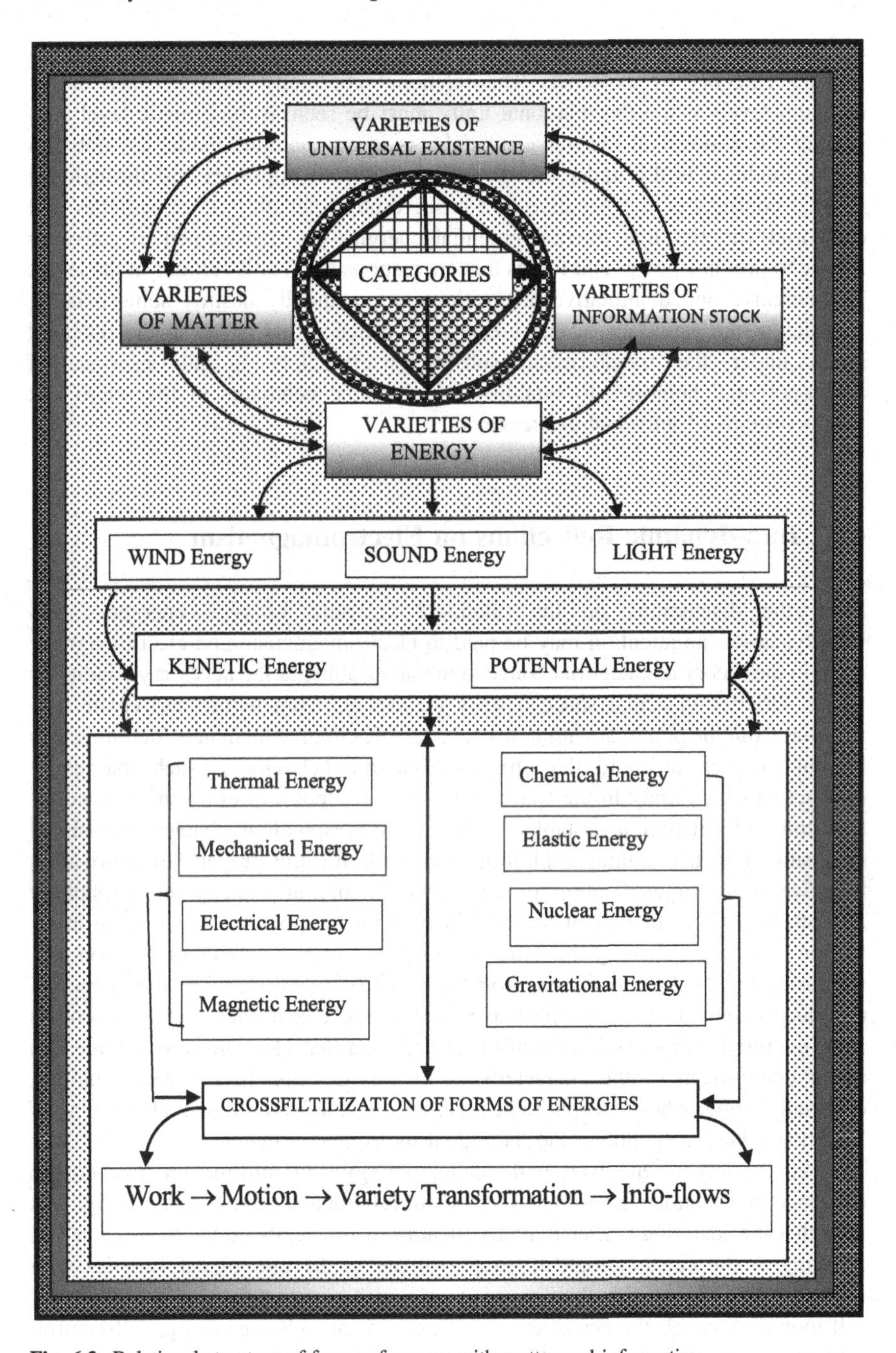

Fig. 6.2 Relational structure of forms of energy with matter and information

and unity among atomic nuclei and atoms. Energy resides in matter which is its primary category of existence. The relational continuity must be seen at the quantum levels while the relational unity must be seen at the organic level supporting the notion that the universal order is in an analogue mode and not in a digital mode. It is the under the general principle of opposites and through the internal conflicts of the negative and positive characteristic sets of varieties and categorial varieties that internal potential energies are transformed to active or kinetic to produce power and forces that are responsible for all motions at the levels of qualitative and quantitative dispositions to continually destroy some existing varieties in the actual space and create new varieties that reside in the potential space under the transformation-substitution principle in all socio-natural actual-potential polarities to establish info-dynamic behaviors of the micro and macro systems of universal existence.

6.4 Info-dynamic Reflections on Electromagnetism

From the discussions on the relational structure between energetics and info-dynamics, an attention may be paid to electromagnetism and electromagnetic force in the theory of categorial conversion that establishes the necessary conditions where varieties in actual space destroy themselves from within and create new varieties from the potential with new characteristics to give them new identities. It is helpful to keep in mind that the info-dynamic behavior is such that every destruction of a variety in the space of the actual is not a destruction of its corresponding information as established by the corresponding characteristic-signal disposition thus maintaining the information stock. It is this property of information that gives a logical justification for the study of history, archeology geomorphology and others. Every creation of a new variety from the potential space is an information flow established by the corresponding characteristic-signal disposition thus increasing the stock of available information at any given time. The importance of the electromagnetic force in the understanding the categorial-conversion process finds expression in the scientifically accepted idea that electromagnetic force is a major determinant of the characteristics of varieties and hence their identities, especially matter whose morphological varieties result from differential interplay of intermolecular forces among the individual molecules in matter.

Every variety is endowed with matter, energy and information to claim its existence and identity from other varieties in relational continuum and unity. It is here that the idea that matter is under plenum of forces also finds expression and meaning in the epistemological space to relate to the ontological varieties with corresponding identities in their existence. It is also here that the concept of internal self-motion, through the activities of self-excitement, self-organizing, self-control, self-teaching, self-learning, self-informing, self-correction, self-destruction and self-creation finds expressions and meaning, where matter is under four continually operating fundamental forces of electromagnetic force, weak nuclear force, strong nuclear force and gravitational force. These four fundamental forces are said to belong to the primary category of forces of matter, in the sense that all other forces belong to the derived category of forces of matter which can be related to the micro-level and the super micro-level, the quantum and the super-quantum levels.

The primary and derived categories of forces relate to the primary and derived categories of energy, both of which relate to the primary and derived category of work in transformations of varieties and categorial varieties. How does one know the difference between a primary category and a derived category and differences among varieties and categorial varieties within the family of categories? In general and in all forms over the epistemological space, the distinction between the primary category and derived categories is the *identification problem* in all knowledge systems as well as the decision-choice systems. It is attempts to find reasonable answer to this fundamental question of distinctions structured in the identification problem in knowing that the claim of information as a property of matter and ontological objects have a convincing justification. Matter is not only a form of energy, it is also a form of information.

The universal existence is thus composed of matter, energy and information as its properties. Energy is also a form of information. It is the information contents of matter and energy that present their commonness and difference of the two categories of existence. The information contents are presented as a unified positive and negative characteristic subsets in terms of characteristic dispositions at the ontological space and projected onto the epistemological space as signal dispositions for acquaintances to reveal different identities of matter and energy as well as various varieties of matter and energy. Nothing is knowable at the epistemological space without information and nothing can be communicated within any source-destination duality without the content component of information. What is a message without a content? Everything transmitted is sensed in varieties with differential content parts of information, where the transmissions are through the signal dispositions under energy.

At any point of time, $t \in \mathbb{T}$, the relational structure of the fundamental category of energy and the characteristic dispositions fix the info-statics of all varieties of matter and energy, with matter and energy identities such that their essences may be known through the distribution of their signal dispositions. The stabilities of the identities and the distribution of the identities of all varieties at any time point depend fully on the relational stability of the interactive behaviors of the elements in the primary category of energy and the distribution of the characteristic dispositions. In respect of this, any time point is a candidate to define the structure of info-statics of varieties and hence any time point may be taken to initialize the study of information production in a forward telescopic process to discover what would be or to claim what would be as knowledge (forecasting and prescription). The same time point may be taken to initialize the study of the past history of the information production in a backward telescopic process to discover what was or the validity of what is being claimed as knowledge.

The backward or the forward telescopic processes encompass a number of epistemic activities including archeological studies, geomorphological studies, the updating of the periodic table, engineering, planning, elements of prescriptive science socio-natural history and others. It is the nature of time and transformations of varieties and categorial varieties, that the study of qualitative and quantitative dynamics requires an indication of the initial information at the selected initial time. It is useful to remember that if $y_t \in \mathbb{Y}$ is a set of time positions taken by quantitative variable y over a time set $t \in \mathbb{T}$ then $(y_t, \ldots, y_{t+i}) \in \mathbb{Y}$ are different time-varieties of the same quantity in a forward telescopic view and $(y_{t-i}, \ldots, y_t) \in \mathbb{Y}$ is also different time varieties in the backward telescopic view for any given quality. Similarly, if $z_t \in \mathbb{Z}$ is a set of time positions taken by qualitative variable z over a time set $t \in \mathbb{T}$ then $(z_t, \ldots, z_{t+i}) \in \mathbb{Z}$ are different time-varieties of the same quality in a forward telescopic view and $(z_{t-i}, \ldots, z_t) \in \mathbb{Z}$ is also different time varieties in the backward telescopic view for any given quality. Any particle in a quantitative motion or qualitative motion is viewed and studied in terms of quantitative or qualitative varieties under intra-categorial or inter-categorial conversions respectively.

Given an arbitrary selected initial epistemic time, $t_0 \in \mathbb{T}$, the forward telescopic process of any variety or categorial varieties may be viewed within the subset of form $(t_0, t_i) \in \mathbb{T} = (-\infty, +\infty)$ where the interplay of the dynamics of the elements of the primary category of energy and force with the characteristic dispositions of varieties will define and establish the info-dynamic field through the destruction of existing varieties in the actual pole and the creation of new varieties in the potential pole in the space of socio-natural polarities. Keep in mind that, generally, the theory of info-dynamics is devoted to the search for the *fundamental principles* that will help to provide the nature of different relational structures of matter, energy and information as forms of energy are generated from within varieties of matter, and actual-ontological varieties are destroyed and new potential-ontological varieties are created under the transformation-substitution principle with real *cost-benefit rationality* in the family of the actual-potential polarities.

The potential ontological varieties include the transformations of varieties over the epistemological space. In these relational structures, matter is transformed into energy which then acts on matter and transform it into matter in stock-flow equilibrium and into continual info-dynamic stock-flow disequilibrium. The energy-matter stock-flow equilibrium ensures the permanency of the universe in relative stability, where the info-dynamic stock-flow disequilibrium ensures continual transformations in construction-destruction polarities, where new varieties emerge from old varieties to maintain the universal stability with a continual information accumulation into the infinite future. There is no expansion in the universe but a continual expansion of space of varieties, family of categories, their identities and the corresponding information set. The information stock-flow disequilibrium provides the cognitive agents the conditions to access the past and bring it to the present and project the present into the future under the *sankofa-anoma* principle of *past-present-future connectivity* in information production and transformation decision-choice activities in the space of varieties.

6.5 A Simple Info-dynamic Reflection on the Einsteinian Fundamental Equation in Physics: An Example of Information Representation

The matter-energy stock-flow equilibria and the info-dynamic stock-flow disequilibrium seem to offer an epistemic path of fundamental conjecture to examine the Einsteinian fundamental formula of physics, where the potential energy $\mathbb{E}$ is related to a mass of a variety $\mathfrak{m}$ and velocity C of electromagnetic activity in a vacuum, where $\mathbb{E} = \mathfrak{m}C^2$. All other equations, whether fundamental or non-fundamental may be examined in the same way. There is no explicit information contents in this formula. This formula, however, can be related to the contents of information defined as characteristic disposition. The energy potential has information stock. The mass of variety has information stock and the electromagnetic velocity is an information flow. To transform this formula into a formula of information phenomenon, one must answer a number of questions. (1) What are the information contents of $\mathbb{E}$? (2) What are the information contents of $\mathfrak{m}$? And (3) what are the information contents of C?

Every mathematical and non-mathematical statement of an idea is an information representation of thought as a message. Such an information representation is either info-stock or info-flow or a combination of both that presents a complexity in knowing. The complexity lies in the epistemic structure that constitutes a potential explanatory justification to the construct of the mathematical representation of an existing phenomenon in the actual space through the actions of acquaintance. The epistemic structure may also constitute a potential prescriptive justification of future phenomenon to be actualized from the potential space. In this case of Einsteinian equation $\mathbb{E}$, the potential energy is revealed by info-stock. Similarly, the mass of the variety $\mathfrak{m}$ is revealed by info-stock. The velocity of the electromagnetic activity in a vacuum C is revealed by info-flow. The formula is thus described by info-stock-flow relation that may be written as $\mathbb{Z}^*_{\mathbb{E}} = \mathbb{Z}^*_{\mathfrak{m}}(\mathbb{Z}_C)^2$, where $\mathbb{Z}^{*\prime}$s are info-stocks and the $\mathbb{Z}$ is info-flow all of which are defined by characteristic-signal dispositions. The electromagnetic activities in $\mathfrak{m}$ will alter the info-stock $\mathbb{Z}^*_{\mathfrak{m}}$ and hence $\mathbb{Z}^*_{\mathbb{E}}$ given the characteristic disposition of the electromagnetic velocity $\mathbb{Z}_C$.

In terms of information, matter and energy relation, one may take the advantage of some accepted ideas in physics in that all matter is composed of atoms which are made up of protons with positive characteristics, electrons made up with negative characteristics and in between them there are the neutrons with some form of neutral characteristics creating atomic relational structures. It is this nature of the atomic relational structure in qualitative and quantitative dispositions that establishes the distribution of the characteristic dispositions of the universal varieties in terms of categorial similarities, differences and identities.

Varieties are eithers internally destroyed or produced as the results of different arrangements and rearrangements of some or all of a number of internal characteristics to change the sequential order of the relational positions. By changing the sequence of these arrangements relative to the number of characteristics and the negative-positive characteristics for the internal and external existence, the existing varieties are destroyed to give way to new varieties and thus alter the electro-magnetic activities and speed and hence the potential energy stock. In terms of information process, the information on the destroyed or transformed the existing varieties are not destroyed but accumulate as info-stock while the information on the newly created varieties acts as info-flows to update the general info-stock. For example, the numbers of protons, electrons and neutrons together establish the quantitative disposition of objects while the sequences of arrangements establish the qualitative dispositions of objects. The combinational structure of qualitative and quantitative dispositions establishes a characteristic disposition and hence a particular variety with a corresponding identity contained in the corresponding characteristic-signal disposition that presents its information of ontological objects.

Each variety and categorial varieties present information flows while the history of their evolution from any time point to any other time point presents a distribution of the information stocks (info-stocks) which is generated by the nature of accumulated distribution of characteristic dispositions. The study of information flows is the study of new varieties and their distribution. It is, thus, an epistemic approach to the study of dynamics of evolving past-present-future of history of universe of things in varieties and categorial varieties. The study of information stocks is the study of existing and non-existing actual-potential varieties and their distributions as seen in their historical past-present-future enveloping. The study of the nature of info-stock is the study of socio-natural conditions at a point of time and the transformation dynamics of socio-natural existence in the space of socio-natural qualitative and quantitative dispositions of varieties in relational continuum and unity. The behavior of the stock-flow information structure over a given time domain is more powerful than it is accorded over the epistemic space. The theoretical structure comes under the general theory of transformation which is divided into sub-theory of categorial conversion and the sub-theory of Philosophical Consciencism. The theory of categorial conversion relates to the necessary conditions which define the *socio-natural necessity* for transformations of varieties [R17.15]. The theory of philosophical Consciencism relates to the sufficient conditions which define the *socio-natural freedom* in the decision-choice space to effect transformations of varieties [R17.16]. Over the epistemological space the necessity relates to the set of necessary conditions which is not the creation of cognitive agents for transformation of each variety. Freedom relates to the set of sufficient conditions which is under the power of cognitive agents to transform varieties where the decision-choice power is constrained by the conditions of necessity under decision-choice intentionality. It is this necessity-freedom relational structure the all forms of engineering find their challenges, the decision-choice action of which lead to failure to transform and unintended consequences.

The epistemic structure of these theories is general for the study of all quantitative and qualitative transformations powered by quantitative and qualitative equations of motion respectively. One thing that must be noted is that the universal existence is nothing but a collection of interdependent varieties in continuum and unity, where the dynamics of such collection is an information stock-flow interdependent. Nothing makes sense over the epistemological space without this relational information stock-flow interdependency. The info-stock-flow conditions of varieties are generated by the possibility of either destructions and the creations, duplications elsewhere or their reconstructions in the future when such information stocks are discovered to reveal the number of elements, the structure of the sequence of their combinations and the techniques and the methods of their formation. It is here that explanatory and prescriptive sciences find utility over the epistemological space in which cognitive agents operate with knowledge production and decision-choice actions on varieties under the cost-benefit rationality. It is also here that engineering of all forms and social institutional engineering find an epistemic residence.

Given the information stock of any variety, the number of elements, that it contains, the sequence of their combination, the techniques and methods of their combinations, the mimicry of the techniques and methods of their formulation are referred to in our contemporary times as technology and engineering where the study of the art and science of the mimicry is the engineering and technological sciences. It is within this epistemic structure, that one can discover a general theory of unity of sciences, a general theory of unity of engineering sciences and a general theory of technological sciences and their dependence on information connectivity with unity of all sciences.

6.6 Information Stoc-Flow Conditions and Languages

The science of all languages with their differences and similarities finds an epistemic meaning, constructability and reducibility under the fundamental principle of varieties. The concept of variety is central to all languages (ordinary and abstract), communication, cognitive understanding, decision-choice actions, human creativity, conflicts, war-peace processes, development and the very nature of existence. Human language takes the forms of spoken and written structures. The spoken language presents itself as permutation of forms of sounds to establish varieties of words, where the sequential arrangements of forms of sounds give vocabulary, connectivity and meaning. This also holds true for written languages where specific characters are used to establish verbal varieties through sequential arrangements of permutation of the acceptable characters to create characteristic disposition of vocabulary of a particular language. Every language is a family of codes in sounds and vocabulary governed by laws of combination to create meaning under variety of vocal sentences. The written languages are representational varieties of the sounds embodied in the words of the languages with strokes called alphabets, where the words are governed by laws of combination to create meaning under varieties of written sentences. In this respect, music is a vocal language with its written component. In the same language, the written characters must have relational connectivity to the sound characters to establish an important isomorphism between the spoken and the written structures of the same language.

Each language presents itself as a sequential arrangements of codes of varieties such that the language contains information defined as characteristic-signal disposition, the meaning of which is revealed by decoding the language codes in the source-destination duality. In other words, a language is encoding-decoding instrument of message storage and carrying information within the source-destination processes over the epistemological space. It may be kept in mind that the contents of any information reside in the characteristic disposition and the messages of any information reside in the signal disposition. All codes are in varieties, so also the signals that carry them. The permutations of these codes create forms of varieties and a family of varieties for identities, categories and recognition over the space of languages. The study of any language is the study of the structure of coding and decoding of varieties of the language in spoken and written spaces. By the use of permutation of these codes and family of codes, thoughts are created and communicated in a particular language and in a manner that establishes varieties of systems of thinking and meaning in the framework of cognitive ingenuity.

Thought, spoken or written in any language, is a message carried by energy through the source-destination transmission or communication process. The system of messaging is a family of systems of information stock-flow dynamics, where written languages are helpful in info-stock, and the spoken language is more helpful in info-flows. It may be observed that message communication in any written language is generally restrictive among all the speakers of the language and available to those that command the technique of the reading and written structure, while the spoken language offers an important flexibility in communication of messages for all irrespective whether one can read or write the language.

The more the codes that are created, and the more complex are the permutations and the family of the permutations, the more complex are the linguistic thoughts and communications. It is here, that languages acquire living characters within the info-statics and info-dynamics, where the expansion of each language is governed by an epistemic process of acquaintance under the governance of info-statics and info-dynamics. The sense here is that the vocabulary of any language can expand within the dynamics of the destruction-creation duality of varieties and the corresponding information under the principle of acquaintance. It is through the general theory of information that the development of the theory of coding acquires scientific significance and analytical usefulness in the theory of knowledge. It is also here that mathematics acquires the character of a universal language, where the varieties of codes can be expanded in relation to thought, and where the mathematical codes can be combined with other linguistic codes of varieties to expand ideas, develop more concepts and increase the space of computability and exactness. It is also here that one finds an increasing understanding of the structure of the family of ordinary languages (FOL) and the terrain that the FOL's cover in thought, communication and decision-choice systems of actions and responses in cognitive existence with the universal environment as viewed from the epistemological space.

6.6.1 *Information and the Family of Ordinary Languages (FOL)*

The concepts (theories) of info-statics and info-dynamics in relations to characteristic-signal disposition, variety, and identity are such that information exists as states and processes in a manner that is independent of ontological objects. In respect of this, information just like matter and energy precedes the formation of languages. A number of questions arise. What is the relationship between information and language in the transmission or communication process? Is information a derivative from a language or a language a derivative from information? Is a language a carrier of information or is information a carrier of language? When information is seen as a derivative from a language then language becomes the primary category for the existence of information, and language must be seen as a property of matter and energy. Language then becomes a driving force of information existence and the continual creation of information and hence varieties, characteristic-signal dispositions, where info-stocks and info-flows become language determined by the conditional existence of some ontological object or objects. In this way, cognitive agents, through language development creates a universal reality and the size of the universe is language-determined. Another question arises as to whether a language is a construct of the mind or the mind is a construct of a language. If language is a construct of the mind then one comes to the classical claim that realities as presented by information are mental images and information in stock-flow dynamics are mentally determined.

On the other hand, when a language is seen as a derivative from information then information becomes a primary category for the existence of a language and information must be seen as a property of matter as is being maintained in this monograph and the companion monograph [R17.17]. In respect of this, information becomes the driving force of the existence of language, and the continual expansion of language, thought, and representation is thus determined by varieties, characteristic-signal disposition and information stock-flow dynamics. Here, perceptions and epistemic constructs are model representations of ontological reality over the epistemological space and not the other way round. In this epistemic frame, any language is a vehicle of information representation and communication that act on the characteristic-signal dispositions of varieties and categorial varieties. Since languages are cognitive representations for message communication, they can be used by cognitive agents to represent misinformation, disinformation and propaganda under the principles of decision-choice intentionality over the epistemological space. This is the misrepresentation problem in information communication over the epistemological space. The misrepresentation problem is an attribute of all epistemological signal dispositions to misrepresent the characteristic-signal disposition in information communication within the source-destination systems under the principle of duality with a relational continuum and unity. Here, every source does not only have a destination support but is a destination, and every destination does not only have a source support but is also a source.

From this conceptual frame, a number of interesting ideas emerge. Differences in the family of ordinary languages are simply differences in sound codes, combination of the codes and symbolic interpretations of their results relative to their information contents, where thought is a representation of message in the source-destination process. The translation of a thought from one language to another is a translation of codes of sounds and symbols of messages to establish similarity of the linguistic varieties and the family of varieties that give an *epistemic isomorphism* to the body of thought in different languages. The use of adjective and adverbs in the FOL's offers an increasing framework of creating linguistic varieties by combining quantitative and qualitative dispositions in an exact-inexact duality for representations and communications in thought and intra-categorial conversions over the epistemological space. It may be kept in mind that linguistic meanings find expressions in varieties. It is through the utility of varieties that linguistic opposites such as antonyms are constructed and also through the utility of categorial varieties that linguistic similarities such as synonyms are constructed to widen the space of linguistic varieties and categorial varieties for source-destination communications over the epistemological space. Antonyms are inter-categorial representations and synonyms are intra-categorial representations.

The expansion of the vocabulary in FOL is driven by the expansion of acquaintance with unknown stock of varieties and categorial varieties as well as the creation of new varieties in terms of info-flows. The greatest characteristic of any FOL is the acceptance of conditions of vagueness in the vocabulary and the linguistic compositions in thoughts that may produce inexactness with different interpretations of and responses to the messages contained in the linguistic constructs. The important analytical strength of any member of the FOL's is its development to deal simultaneously with quantitative and qualitative disposition from which linguistic quantities, such as big, tall, small, plenty and others are developed as elements of variety along with linguistic qualities such as beauty, colors, pain, suffering and others are also developed as elements of varieties. Such a linguistic development allows the expansion of the space of varieties by increasing the permutation of quantitative and qualitative varieties, where an increasing complexity creates increasing forms of vagueness which require subjective interpretations of information contained in the messages carried by the linguistic structures. The same increasing complexity creates increasing forms of cognitive limitations which require objective interpretation of information contained in the messages carried by the linguistic structures.

In respect of this epistemic frame, more vocabularies are created in terms of synonyms and antonyms for comparative linguistic varieties and categorial varieties that can accommodate the messaging system of info-statics and info-dynamics associated with the comparative ontological varieties and categorial varieties. The comparative varieties and categorial varieties with cognitive limitations and the linguistic vagueness over the space of acquaintances are the results of universal continuum and unity that FOL's must deal with from the ontological information to the construct of epistemological information. Indeed, it is here that the development of either dictionary, a thesaurus or glossary of terms enters as useful to definitions of concepts, words, ideas and others as well as useful to explications in science and knowledge systems. The concept and development of a dictionary or thesaurus are impossible under conditions of complete linguistic exactness in words and phraseology. Vagueness is both a benefit and a cost in representation and communication in FOL's where linguistic exactness may form an important constraint in the linguistic expansions and cultural evolution. The stock of vocabulary of any language relates to the known info-stock of information, while the increases in the vocabulary in any language relates to the known info-flows of universal existence over the epistemological space.

6.6.2 Information and the Family of Abstract Languages

The presence of vagueness in the FOL's for information representation and communication creating possibilities of different interpretive conditions of the information contained in messages in the exact-inexact duality presents some important problems and difficulties in all scientific works and messaging systems seeking to limit disagreements. The vagueness while amplifying the interpretive region of human linguistic information relations, creates penumbral regions which constrain scientific exactness in decision-choice systems in transformations of varieties and categorial varieties over the epistemological space. Usefully, the existence of the family of ordinary languages and the understanding of the structure of their developments provide an important cognitive space for the creation and development of other forms of languages that are not ordinary but abstract in order to reduce the problem of vagueness in the linguistic construct of representation, and the communication of information contents within the source-destination duality with relational continuum and unity. It may be kept in mind that the concept of the source-destination duality with relational continuum and unity implies the concept of categorial reversal of the actors, where the source is also a destination and the destination is also a source and that every source has a corresponding destination and every destination has a corresponding source in the system of messaging processes. In this epistemic process, how does one view the concept of an abstract language, the family of abstract languages (FAL) and the possible size of the family that may be developed?

In thinking about these relevant and important concepts in the general theory of information, a view of the concept of abstract language may be seen from the framework of mathematics and its development. Mathematics is seen as an abstract language developed in the framework that allows its simultaneous usage with any element in the FOL's with proper instruction. The mathematical approach to language formation and development are to solve the problems of vagueness in representation of information contents and its communication when used in the FOL's. The requirements of exactness in mathematical representation is an important motivation in the development of the theory of measurements wherever possible in all areas of knowledge production. These were extended to solve the problem of signal-disposition limitation through acquaintance. It may be kept in mind that every element in the communication processes, broadly defined, has quantitative and qualitative characteristics which are divided into negative and positive characteristics. The family of ordinary languages (FOL) is to deal with the representation and transformation of the quantitative and qualitative dispositions in all communications. To accomplish these tasks the FOL must deal simultaneously with semantic and syntactic dispositions to account for vagueness in inexactness-exactness duality in all communications that must transmit quantitative and qualitative information contents, individually and simultaneously, through the messaging systems under conditions of subjective-objective phenomenon.

Things are different in the development of the mathematical language and other forms in the family of abstract languages (FAL). The main task of the development of any element in FAL is to reduce or eliminate the presence of vagueness associated with representation and transmission of information in FOL. As much attention is concentrated on the representation and transmission of quantitative-disposition component of information contents, where an exactness is the ideal goal sought to the neglect of the qualitative-disposition component of information contents in the messaging system. Toward this goal of exactification, the elements in the FAL's are devoted to the development of perfect syntactic specification in representation and transmission of abstract disposition which is a set of non-real objects such that their manipulation under specified rules within the abstract language are independent of considerations of semantic conditions of transmission of information contents in the source-destination duality. In other words the FAL's exist in context-free domain. Does this mean that FAL's exist in content-free environment?

From the viewpoint of the general theory of information, composed of the theory of info-statics and the theory of info-dynamics, what types of information are represented and transmitted or be represented and transmitted by any member in the FAL's? What does it mean in the information theory to speak of an abstract language and the corresponding family and possible expansion? In what manner do the elemental members in FAL's deal with quantitative and qualitative dispositions in time and over time where in time refers to info-statics and over time refers to info-dynamics? How is any member of the FAL's used as a vehicle to communicate among cognitive agents? Are the members in the FAL's developed as a means to communicate among abstract machines and to facilitate the works of axiomatic control systems? Are the FAL's intended to facilitate inter-personal relationships or person-machine relationships or machine-machine relationships? Are FAL's programming languages? What are the relationships between abstract language and information? How does one distinguish any language from another, for example the differences and similarities within the FAL's, and between the FOL's and FAL's or between English and Chinese? The answers to these question come from the information concept of variety and corresponding identity as presented by the distributive nature of characteristic-signal distributions to reveal the differences and similarities in their contents.

Epilogue and Conclusions

The info-dynamics is in relation to the production of information over both the ontological and epistemological spaces. The analytical framework identifies the dynamics of ontological information and the dynamics of epistemological information. The epistemological info-dynamics and ontological info-dynamics have similarities and differences. The understanding of these differences and similarities is essential in understanding the concepts and actions of information transmission, information communication, representation and language. In the development of the theories if info-statics and info-dynamics and the understanding of what are to be accomplished it is necessary to distinguish among the intra-ontological information relations, intra-epistemological information relations and inter-ontological-epistemological information relations. The intra-ontological information relations and the inter-ontological-epistemological relations are maintained among ontological objects through the actions of transmission. The intra-epistemological information relations among epistemological object are maintained through actions of communication. Notice that when we speak of climate change and environmental decay, we are speaking of the ontological response to epistemological transmission of information. In epistemic simplicity, all lives exist as information relations to things in the problem-solution space under complex decision-choice actions, where a solution to a problem generates a new problem. The problem-solution process is information-relation determined in never-ending process of existence defined by info-stock-flow processes. The multiplicity of information dynamics are decision-choice determined producing new info-flows and updating info-stocks. The study of info-dynamics is also the study of decision-choice dynamics.

A Reflection on the Ontological and Epistemological Information Structures

The conclusions in this monograph are used to summarize some important points in the theory of info-dynamics relative to the theory of info-statics. The epilogue is used to reflect on the methodological difficulties, analysis and directions for the future research to create foundations for the reexamination of the concept of entropy

K.K. Dompere, *The Theory of Info-Dynamics: Rational Foundations of Information-Knowledge Dynamics*, Studies in Systems, Decision and Control 114, https://doi.org/10.1007/978-3-319-63853-9

in the framework of general information-knowledge system in the source-destination duality with a relational continuum and unity.

The accepted set of initial ontological conditions must be defined by a characteristic disposition which is then transmitted through the corresponding signal disposition for the construct of epistemological information. The constructed epistemological information will provide a framework for all epistemological activities including ordinary and abstract representations, knowledge-production processes, abstract universal transformation, cognitive activities, and decision-choice actions. In all analytical framework, the differences and similarities of ontological information and epistemological information must be clearly understood. The understanding will affect our perceptions of information transformation and information communication and the areas of utilization that may be required of them as well as limitations that may be imposed on cognitive capacity in epistemic actions in relation to the zone of irreducible stupefied ignorance.

The conclusions of the monograph finds expressions in a theoretical acceptance of a fundamental framework for dynamic analysis irrespective of the area of general information-knowledge system that cognitive agents may pursue in reach, teaching, learning and decision-choice actions. The theoretical construct of info-dynamics presents a logically general framework for building a general theory of unified sciences in static conditions and dynamic process of transformation in destructive-constructive mode on the basis of information to explain socio-natural forces of change of varieties in individual and collective domains where all socio-natural elements of states and processes are viewed in terms of varieties, categorial varieties and variety transformations to maintain the notion that universal permanency finds residence in change. In social setting, different socioeconomic institutions are also viewed in terms of varieties, where the existing institutional varieties are destroyed from within under their own yoke of inefficiencies in dealing with new problems arising from the changing micro-varieties. Simultaneously with the dying of the existing institutions, new institutions are created from within with the ashes of the old ones as the inputs in their replacements under the general principle of the opposites to deal with conditions of new problems which emerge from the solutions to the old problems by the existing institutions and dynamics of the micro-structures within the social institutions. Under the principle of opposites, every variety exists as a dualistic negative-positive relational structure, in continuum and unity, which provides its identity, essence and dynamism under continual conflicts and conflict resolutions.

There are two information subspaces that define the organic information subspace. They are the ontological information subspace that holds the ontological information at every point of time and the epistemological information subspace which holds the epistemological information at every point of time as have been previously established in this monograph and in the theory of info-statics [R17.17]. The epistemological information is contained in the ontological information, the expansion of which is constrained by important cognitive limitations to generate all kinds of uncertainty that further divide the epistemological information into sub-sectors which are presented as a cognitive geometry in Fig. 1. The uniqueness of the

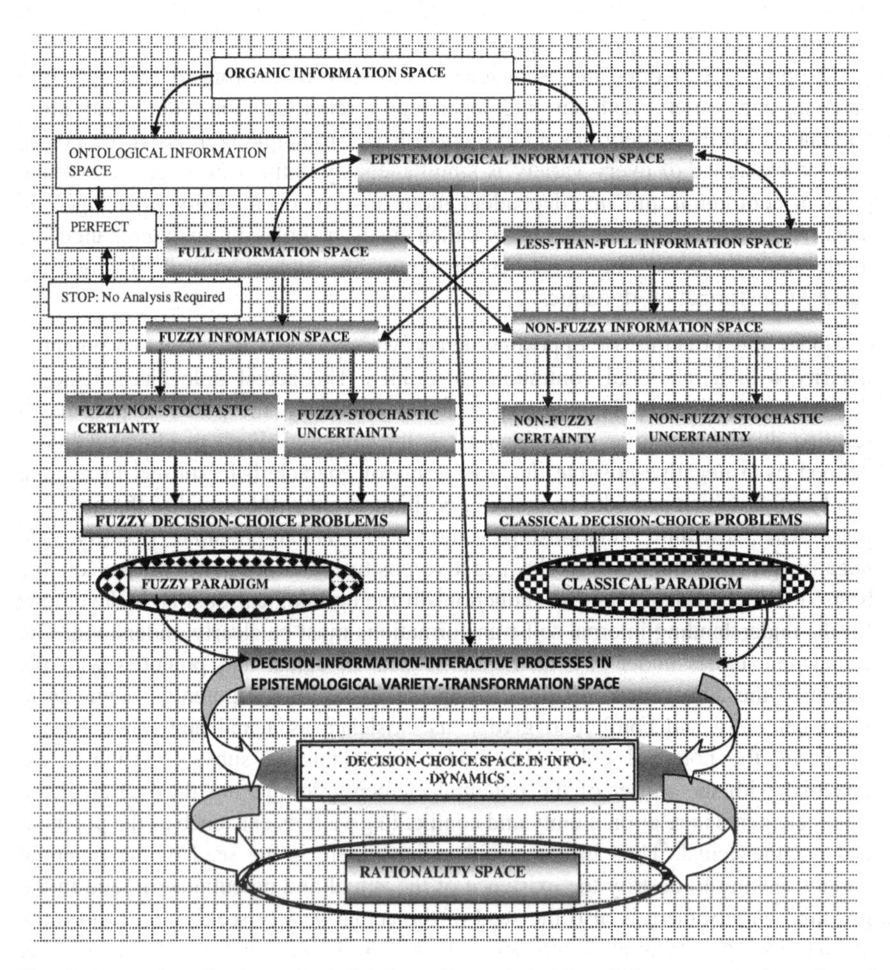

Fig. 1 Categories of epistemological information and decision-choice processes in static and dynamic spaces

ontological information is its perfection in relation to ontological info-processes. The epistemological information is distinguished from the ontological information in terms of defectiveness in relation to info-processes.

The Principle of Opposites in Transformations

The transformation of varieties from within and the maintenance of sustainability of continual change are explainable through the theoretical foundations of *the principle of opposites*.

Definition (*The Principle of Opposites*)

The principle of opposites is in relations to state of existing of two opposing parts in one variety with an interdependently relational continuum and unity without which the variety has no existence. It is a principle that assert the notion that the existence of every variety is simultaneously defined by negative-positive parts in continuum and interdependent mode to provide its unity of being. The negative-positive parts are opposites called duals or poles in an interchangeable mode and projected into cost-benefit space in support of all input-output variety-transformation dynamics.

Under the principle of opposites, every variety exists as an actual-potential polarity with a relational continuum and unity. The actual pole has a residing duality which gives its identity as the actual. Similarly, the potential pole has a residing duality that gives it its identity as the potential. Each residing duality is defined by a negative-positive duals in terms of characteristic subsets in relative combination which provides its residing position in either the actual or the potential pole. The structural relationships among the actual-potential polarity, the negative-positive duality and the negative-positive characteristic subsets on the basis of which constructionism-reductionism methodological duality is developed for the understanding internal self-transformations must be clearly understood. The actual pole exists under two elemental subsets of actual and potential with the subset of the actual elements dominating the subset of the potential elements in relational unity and conflict to give it its existence as actual and not potential. Similarly, the potential pole exists under two elemental subsets of actual and potential with the subset of the potential elements dominating the subset of the actual elements in relational unity and conflict to give it its existence as the potential and not the actual. Thus in the actual variety, there is the seed of the potential variety seeking to destroy the existence of the actual and potentialize it, and in the potential variety there is also the seed of the actual variety seeking to destroy the existence of the potential variety and actualize it. These activities of actualization and potentialization are made possible by the nature of the residing dualities in terms of negative and positive characteristic subsets that define the relative composition of the actual-potential combination in each pole.

Any duality is said to be negative if the negative dual dominates the positive dual in characteristic subsets. It is said to be positive if the positive dual dominates the negative dual in of characteristic subsets. Each duality exists in conflict under relational continuum and unity in an inter-supportive structure. This dualistic representation is extended to the actual-potential polarity which may also be represented as negative-positive polarity or the actual-potential structure may be extended to the representation of duality and the corresponding duals. The actual pole of the actual-potential polarity may be viewed as the positive pole while the potential pole may be viewed as the negative pole. In other words, the actual-potential polarity may be viewed as the positive-negative polarity for analytical convenience of definition, explication and exposition in the methodological constructionism in the exposition of rightward logical structure from the primary category to the derived categories of universal existence, and in the methodological reductionism in exposition of leftward logical structure from the derived categories to the primary category of universal existence.

The Relational Structure Between the Polarity and Duality

There is an important relational structure between the infinite set of polarities and the infinite set of dualities. The distribution of the relational structures of the dualities and polarities will indicate the socio-natural processes and processors with the corresponding technologies for the variety transformations to effect info-flows. The relational structure is such that in the actual pole resides the positive duality, while the negative duality resides in the potential pole of the polarity where every polarity is in relational continuum and unity with conflicts that seem to show themselves as contradictions. The epistemic structure is a representation of a system of dualistic actual-potential polarities that will allow a theoretical explanation of info-dynamics through variety transformations. The structure is such that every positive duality in the actual pole has a supporting negative duality in the potential pole to ensure their mutual existence in a temporary mode, and mutual destruction for a continual disequilibrium process. The positive duality in the actual pole is composed of the dominant positive dual with a mutually supporting existence with the dominated negative dual for the game of destruction of the actual and its conversion to the potential through the internal mutual destruction, creation and negation under the internal matter-energy activities with decision-choice actions of the dualities. Similarly, the negative duality of the potential pole is composed of the dominant negative dual with a mutually supporting existence with the dominated positive dual for the game of destruction of the potential pole and its conversion to the actual through the internal mutual destruction, creation and negation under matter-energy activities with decision-choice actions of the dualities.

The residing dualities are the game instruments for the internal transforming decision-choice actions of any actual-potential polarity, where the actual is destroyed from within and sent to the potential space with complete reservation of its corresponding information. The positive internal duality becomes the negative internal duality with the reverse of the relational dualistic dominance in the game of variety transformations. In its place, a potential pole is created from the potential space and actualized in the space of the actual varieties by forces generated by the production of the internal energy through the dualistic negative-positive relational conflicts in continuum and unity, where the negative internal duality becomes the positive internal duality with the reverse of the relational dualistic dominance in the game of variety transformations of existence and stock-flow information conditions. In an essence, the existing actual pole serves as an input (real cost) into the production activities of the existing potential, the output of which is a new actual (real benefit), while the existing potential serves as an input (real cost) into the destructively productive activities of the existing actual the output of which is a new potential (real benefit). The general process of transformation of actual-potential polarity is a real cost-benefit process where the conditions of actual is constrained by the conditions of the potential which is also constrained by the conditions of the actual.

The Cost-Benefit Conditions of Variety Transformations and the Info-dynamic Process

The input-out relations must be seen in terms of cost and benefit information in all the transformation processes. All varieties and categorial varieties reside in the cost-benefit information relation. The real net benefit of an existing actual variety is a real cost in obtaining the real net benefit of a potential variety. In a dualistic relation, the real net benefit of a potential variety is a real cost in retaining the net benefit of an existing variety. The structure is such that within each duality, the positive dual may be associated with benefit conditions and the negative dual may be associated with the cost conditions in an inter-supportive relation. The relative structure presents the net benefit of existence. In a logical extension, the positive duality has a positive net benefit that provides the actual pole its existence while the negative duality has a negative net benefit that ensures the existence of the potential pole. The existence of a polarity is defined by cost-benefit info-stock in the relativity of poles. The transformation of an actual-potential polarity is defined by cost-benefit info-flow to update the info-stock. In this way, every characteristic of variety exists as mutually inter-supportive cost-benefit information relation in the maintenance of the respective existence of the duality and polarity.

In the general epistemic framework of static and dynamic conditions, every pole of the actual-potential polarity resides in the conditions of real cost-benefit information duality to provide its existence and changing existence with real net cost-benefit values in the existentially universal system of being. The conditions of dualistic real net benefits and their active role in the general dynamics of actual-potential polarity must be clearly understood. These conditions form the fundamental essence of, and justification for cost-benefit analysis and the corresponding real cost-benefit rationality that finds epistemic reflections and meaning in the solution to the Asantrofi-anoma problem of universal transformation decision-choice dynamics in the space of info-stock-flow conditions under no-free-lunch principle within all epistemological activities. The Asantrofi-anoma problem is dualistic decision-choice condition under cost-benefit structure where every decision-choice variety simultaneously contains cost and benefit in such a way that one cannot choose the benefit and leave the cost. The actual pole is an opposite relative to the potential pole with its cost-benefit balance for its existence. The potential pole is an opposite relative to the actual pole with its cost-benefit balance for its existence. The actual and the potential poles are connected by the real cost-benefit information conditions, where the actual pole supplements the real cost-benefit conditions which are wanting in the potential pole to make its temporary existence to reside within a complete-incomplete duality, and the potential pole also supplements the real cost-benefit conditions which are wanting in the actual pole to also make its temporary existence to reside within a complete-incomplete duality. In transformation dynamics, the actual and potential poles exist in a complex real cost-benefit of relationally complex interdependencies without which the actual-potential polarity of any variety has no meaningful existence. Each pole also exists relative to the dynamic behavior of the residing duality in relational continuum and unity. The duality with a relational continuum and unity

It has been argued that information is a set of relations among things besides establishing identities of elements as varieties. These identities are made known to other varieties through the relations that the information engenders. The subset of information relations reveals itself as a subset of cost information relations and subset of benefit information relations that are relationally inter-supportive in conditions of reciprocity which involve a relational asymmetry and symmetry in categorial transformations of varieties in a manner that effects the decision-choice activities and the direction of transformations over the epistemological space. The directions of transformations over the ontological space is nature-determined in relation to ontological assessments relative to cost-benefit conditions in terms of the universal designs of the primary existence with variety transformations in the universal environment without uncertainties. The directions of transformations over the epistemological space are under the decision-choice actions of cognitive agents in relation to their relative assessments of incomplete and fuzzy real cost-benefit information conditions under various different scenarios of possible outcomes and their probable effects as seen within the real cost-benefit duality. The ontological information structure is perfect in terms of completeness and exactness and the corresponding transformation outcomes are sure without fuzzy-stochastic conditionality. The epistemological information structure is imperfect in terms of incompleteness and inexactness and the corresponding transformation outcomes are unsure with fuzzy-stochastic conditionality

The Principle of Opposites and the Fuzzy Paradigm of Thought

The information requirements for representation of all principle of opposites over the epistemological space is incomplete and fuzzy on one hand and objective and subjective on the other hand. Such an information relation is referred to as fuzzy-stochastic or stochastic-fuzzy information structure, where fuzziness relates to the vagueness and stochasticity relates to the volume-incompleteness of the available information. This information structure in the in the set of the logical building blocks of the epistemic framework for deriving and explaining the propositions of the theory of info-dynamics is very important to note in terms of the respective roles of the building blocks and the manner in which they work to generate variety transformations. The structure of the principle of opposites has been defined, explicated and logically established in sections II and III, where the opposites of any variety exist in a relational continuum and in an interdependent unity. The properties of the relational continuum and inter-dependent unity impose conditions of existence that requires a specific paradigm of thought which will simultaneously deal with the information conditions of relational continuum and interdependent unity which are required for continual analysis of the opposites in energy, power and force generation through the give-and-take conflicts in the variety transformation games. These variety transformation games are between the poles of polarity on one hand, and the duals of the duality on the other hand. The reward is power and dominance. The information conditions of the principle of opposites are enhanced by subjectivity and objectivity to which the paradigm must also account for in the transformation decision-choice activities. The needed paradigm must be able to support the notion that the poles of any polarity and the duals of any duality are interdependent in complementarity, conflicts and unity in existence.

The required paradigm of thought must have an ability of simultaneously logical information representation of conflict, complementarity and unity of existence of all actual and potential varieties as seen in polarities and dualities. In terms of the principle of opposites, every variety resides in existence-non-existence characteristic conditions under relational conflicts and unity in positive and negative categorial conversions, where existence of any variety is being transformed to non-existence and non-existence of the same variety is being transformed to existence. Existence is the actual and non-existence is the potential. Every variety must be informationally represented as actual-potential opposites that are relationally connected in existence and nonexistence without which its existence and identity are impossible. Within the principle of opposites, simultaneously reside the sets of necessary and sufficient conditions for existence and non-existence, with transformation and non-transformation of the necessary and sufficient conditions of info-stock-flow dynamics. Here, the necessary conditions of varieties are defined by static states and the sufficient conditions for new varieties are established by transformation decision-choice actions to produce info-flows. Under the logical structure of the principle of opposites, the characteristic sets for defining the opposites are mutually non-exclusive and collectively exhaustive to share the environment of their collective existence.

The required paradigm of thought with an ability of simultaneously logical information representation of conflicts, complementarity and unity of existence of all actual and potential varieties as seen in polarities, dualities and all opposites is the fuzzy paradigm. The fuzzy paradigm of thought accept the information representations of logical extremes of existence (truth) and nonexistence (falsehood) and all the sets of pairs of existence and non-existence (truth-falsehood) in proportionality that resides in the existence-nonexistence polarity. In this representation, every information must be represented as a fuzzy set in a dualistic structure to reveal the existence of the opposites of zero and one as extremes with inclusion of all pairs between them.

Definition (*Fuzzy Paradigm*)
The fuzzy paradigm of thought, composed of its information representation, logic and mathematic is a method of knowing that recognizes the methodology of constructionism-reductionism duality in primary and derived categories of all existence. Every proposition, statement, knowledge claim, variety existence or others exists as a set in combination of its opposites seen in relational proportionality. The fundamental principle of information representation is an acceptance of the notion that every existence is made up of negative-positive characteristic subsets in relational continuum and unity. Corresponding to this fundamental information representation is the logical reasoning that holds that all proposition contain true and false characteristics in varying proportions, where the acceptance of true or false proposition is simply by *decision-choice actions* operating on some rationality over the epistemological space. The corresponding mathematics is set-theoretic or categorial approach to the information representation with fuzzy indicator function to construct a family of category of belongings by the method of fuzzy decomposition. The points of decomposition are the transversality points which are obtained by fuzzy-optimization reasoning that allows the construct of either fuzzy conditionality, fuzzy-stochastic conditionality or stochastic-fuzzy conditionality for the qualification of the degree of acceptance (non-rejection) or non-acceptance (rejection) of the proposition.

For example, variety existence is seen as a fuzzy set of degrees of existence $\mathbf{A}$ with a generic element $\mathbf{x} \in \mathbf{A}$ and a membership characteristic function $\mu_{\mathbf{A}}(\mathbf{x}) \in [0, 1]$ where $\mu_{\mathbf{A}}(\mathbf{x}) = 0$ implies complete nonexistence and $\mu_{\mathbf{A}}(\mathbf{x}) = 1$ implies complete existence and $\mu_{\mathbf{A}}(\mathbf{x}) \in (0, 1)$ implies simultaneity of existence and nonexistence in the distribution of relative structural degrees. The set $\mathbf{A}$ may be viewed as the set of existential varieties where each variety has a fuzzy number defined by $\mu_{\mathrm{A}}(\mathbf{x}) \in [0, 1]$. Similarly, in the same space we can specify the opposite as a fuzzy set of non-existential varieties $\mathbf{B}$ with a generic element $\mathbf{x} \in \mathbf{B}$ and a membership characteristic function $\mu_{\mathbf{B}}(\mathbf{x}) \in [0, 1]$ where $\mu_{\mathbf{B}}(\mathbf{x}) = 0$ implies complete existence and $\mu_{\mathbf{B}}(\mathbf{x}) = 1$ implies complete non-existence and $\mu_{\mathbf{B}}(\mathbf{x}) \in (0, 1)$ implies simultaneity of non-existence and existence in the distribution of relative structural degrees. In this dualization, if $\mu_{\mathbf{A}}(\mathbf{x}) = 0$ then $\mu_{\mathbf{B}}(\mathbf{x}) = 1$ and if $\mu_{\mathbf{A}}(\mathbf{x}) = 1$ then $\mu_{\mathbf{B}}(\mathbf{x}) = 0$ to define logical extremes. In the fuzzy paradigm of thought, all varieties simultaneously reside in existence and non-existence in relational continuum and unity in such a way that every proposition has a support of its opposite without which the proposition has no meaning. The system of logical claims is such that in the fuzzy set $(\mathbf{B}, \mu_{\mathbf{B}}(x) \in [0, 1]) \subseteq (\mathbf{A}, \mu_{\mathbf{A}}(\mathbf{x}) \in [0, 1]) \subseteq (\mathbf{B}, \mu_{\mathbf{B}}(\mathbf{x}) \in [0, 1])$ such that the decision on the characteristic set $\mathbf{A}$ is constrained by the characteristic set, $\mathbf{B}$ where one seeks an optimal value in the fuzzy set $(\mathbf{A} \cap \mathbf{B}, \mu_{(\mathbf{A} \cap \mathbf{B})}(\mathbf{x}))$ as the common decomposition boundary of the duals, $\mathbf{A}$ and $\mathbf{B}$ of the duality or the poles of the polarity. The usefulness of the fuzzy analytical approach is the emphasis on the notion that all claims over the epistemological space are decision-choice determined with either fuzzy, stochastic-fuzzy or fuzzy-stochastic conditionality.

The analytical process also bring into focus the objective-constrained logical-mathematical reasoning where a selection is made from the objective set $(\mathbf{A}, \mu_{\mathbf{A}}(\mathbf{x}) \in [0,1])$ constrained by the elements in the constraint set $(\mathbf{B}, \mu_{\mathbf{B}}(\mathbf{x}) \in [0,1])$ and vice versa. The same sets define the real cost-benefit conditions in the variety transformations where each one is either a real cost or real benefit depending on the nature of the variety.

In this epistemic respect, every claimed and unclaimed statement reside in a true-false duality in relational continuum, unity with mutual and existential inter-supportive mode. Translated into universality of varieties and categorial varieties, we have the life-death duality in relational continuum and unity, where life resides in death as its supporting opposite and death resides in life as its supporting mode. Similarly, existence as an actual pole resides in non-existence as the potential pole for its support and nonexistence as a potential pole resides in existence as an actual pole in a relational continuum and unity in such a way that both of them complete the existential universality of the actual-potential polarity. The meanings of the concepts of duality, negative-positive duals, polarity, actual-potential poles, variety and the knowledge about their states and processes them are made available over the epistemic space through epistemological information as the third dimension of universal knowing.

Over the epistemological space, this epistemological information comes to cognitive agents as fuzzy-stochastic information in the sense of vagueness and incompleteness under the principle of acquaintance with the signal dispositions. The representation of vagueness in information was the central debate between Russell [R3.79, R3.81] and Brouwer [R3.14, R3.15, R3.22] that also formed the foundational debate between the intuitionist School and the formalist schools of mathematical development in terms of representation of information and the logic of analysis [R3.6, R3.26]. The representation of incompleteness finds expressions in the stochasticity or probability with exact or inexact representation [R2.17, R2.23, R2.24, R2.32, R2.40]. The concept and the method of exact and complete representation of epistemological information is the position of the formalist with as an artificial imposition in the universal understanding to allow the use of the classical paradigm. This approach creates difficulties in the understanding of transformation processes of varieties where the logic deals with change or no change with complete excluded middle. The use of excluded middle means one cannot meaningfully of the process of change in terms characteristic transformation. The inexact and incomplete epistemological information is the natural characteristic of all characteristic-signal dispositions of varieties and its representation must find meaning in a paradigm of thought that allows inexactness and incompleteness of epistemological information to be represented with logical analysis that accepts continuity and contradictions and rejects excluded middle.

It is here that fuzzy paradigm of thought, composed of its logic and mathematics offers a powerful analytical advantage over the classical paradigm of thought with its logic and mathematics under the conditions of excluded middle where every proposition is a listed set of two elements of the form {0, 1} indicating a membership value of belonging (1) or non-belonging value as (0). In the classical exact representation of belonging and non-belonging, there is no relationality of degrees of belonging for interdependency. With the logic of excluded middle, every proposition exists as a logical point which translates to an existence or nonexistence but not both and the system sits on no-give-and-take mode. This classical paradigm applies to the extreme in a digital form and has a limited application in the understanding of continual qualitative-quantitative variety transformation and the general dynamics of stock-flow conditions of information and knowledge production. The classical paradigm of thought with its excluded middle has an important epistemic difficulty, where considered in proportional structures cannot fulfil its main proposition.

In the fuzzy paradigm of thought, every proposition and knowledge claim over the epistemological space are represented as a set of degrees of belonging in opposites, where truth is proportion to falsity, existence is proportional to nonexistence, and bad is proportional to good and many others without which the meaning of these concepts are vacuous. In this epistemic respect, the claim of truth, existence, good and others are formulated as fuzzy decision problems with the application of methodology of fuzzy optimization of the membership characteristic function of the positive dual subset subject to the membership characteristic function of the negative dual subset in order to abstract the optimal point of demarcation in proportionality of belonging. The point of demarcation is $\mathbf{x}^* \in (\mathbf{A} \cap \mathbf{B})$ such that $\mu_{\mathbf{A} \cap \mathbf{B}}(\mathbf{x}^*) = \mu_{\mathbf{A}}(\mathbf{x}^*) = \mu_{\mathbf{B}}(\mathbf{x}^*) = \alpha \in (0,1)$. This α-point of demarcation is then used to induce a fuzzy decomposition into dualistic dominance in the game of variety transformation within the polarity into actual and potential poles as well as in duality into negative and positive duals that relate to info-stock and info-dynamic structures. The great advantage of the fuzzy mathematical component of the fuzzy paradigm is the availability of a set of mathematical functions that may be used in dualistic representations of fuzzy-stochastic information which allows one to speak of fuzzy-random variable and random-fuzzy variable in fuzzy-stochastic spaces. In general, therefore, the basic constructionism-reductionism of methodological dualism in the understanding the process of continual change is to simultaneously work from the principle of opposite composed of polarity and duality with relational continuum and unity by rejecting the principle of excluded middle, accepting contradictions as valid in thought and existence in combination with the fuzzy paradigm of thought in information representation and analytical construct of knowledge systems. The process of transformation is understood in terms of conflicts and games where dominance is the payoff through which existing varieties are destroyed and transformed into new varieties from the potential space.

Fuzzy Numbers and Fuzzy Characteristic Membership Functions

In the analysis of polarity and duality with relational continuum and unity through the fuzzy paradigm of thought, a number of membership functions dealing with simultaneous representations of the duals and poles are provided in [R4.10, R5.35, R5.13]. These include different info-static and info-dynamic conditions in terms of representation of the duals in duality, the poles in polarity and the respective dualities in respective poles of the polarities of varieties, and in relation to all transformations for the analytical constructs of the theory of info-dynamics in the production of info-flows to update the info-stocks in the universal system of existence and its dynamic behavior in disequilibrium process. For example, the triangular fuzzy number for the degree of existence may be specified graphically as in Fig. 2.

The TFN (triangular fuzzy number) is such that the increasing left upper arrow in ZONE-E indicates an increasing existence. Translated in transformation conditions, the variety is gaining an increasing existence from nonexistence to becoming actualized. The decreasing right lower arrow in ZONE-E indicates a decreasing existence and translated in transformation conditions the variety is losing existence

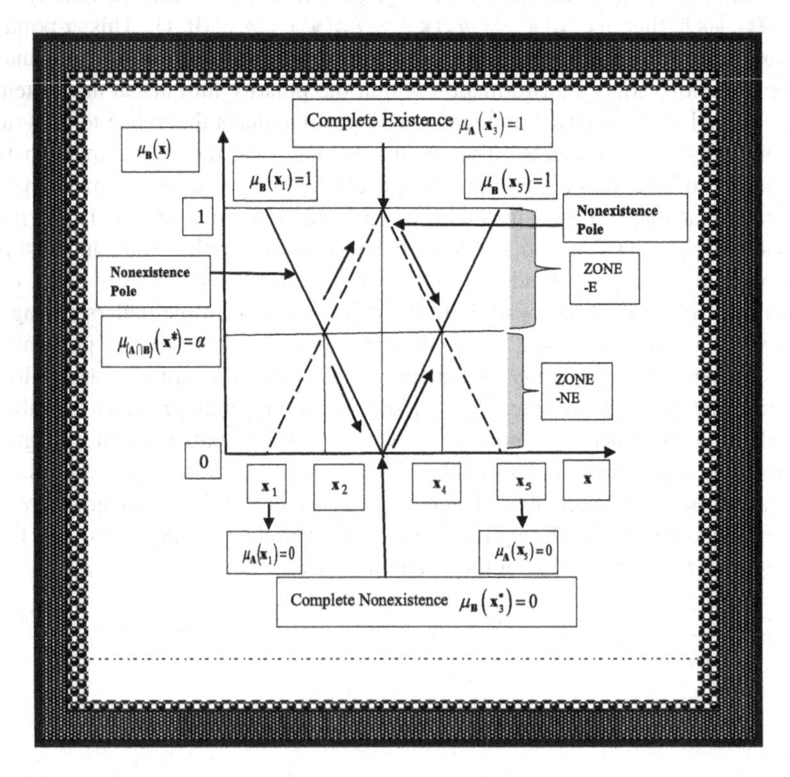

Fig. 2 Fuzzy representation of existence-nonexistence polarity

and taking up nonexistence. In terms of polarity, the any pole in ZONE-E holds dominance over any pole in ZONE-NE which is dominated in the game of transformation. Between $\mathbf{x}_1$ and $\mathbf{x}_2$ as well as between $\mathbf{x}_4$ and $\mathbf{x}_5$ the positive (actual) pole dominates the negative (potential) pole of the polarity in existence and in the transformation process. The points $\mathbf{x}_2$ and $\mathbf{x}_4$ are the breakeven points that define the transversal points or the tipping points in the transformation game where either the actual pole is fading way and the potential pole is emerging or the actualized potential pole is fading away to make room for a new actualized potential. Between $\mathbf{x}_2$ and $\mathbf{x}_4$ the negative (potential) pole of the actual-potential polarity dominates the positive (actual) pole in the variety-transformation process where the potential has been actualized. Analytically, we may specify the TFN for existence-nonexistence polarity as in Eq. (1):

$$\left.\begin{array}{l} \mu_{\mathbf{A}}(\mathbf{x}) = \left\{\begin{array}{l} 1 \text{ if } \mathbf{x} = \mathbf{x}_3^* \\ \in (0,1) \text{ if } \mathbf{x} \in (\mathbf{x}_2, \mathbf{x}_4) \\ 0 \text{ if } \mathbf{x} \leq \mathbf{x}_1 \text{ or } \mathbf{x} \geq \mathbf{x}_4 \end{array}\right\} \text{Potential (Negative) Pole} \\ \mu_{\mathbf{B}}(\mathbf{x}) = \left\{\begin{array}{l} 0 \text{ if } \mathbf{x} = \mathbf{x}_3^* \\ \in (0,1) \text{ if } \mathbf{x} \in (\mathbf{x}_2, \mathbf{x}_4) \\ 1 \text{ if } \mathbf{x} \leq \mathbf{x}_1 \text{ or } \mathbf{x} \geq \mathbf{x}_4 \end{array}\right\} \text{Actual (Positive) Pole} \\ \mu_{\mathbf{A}}(\mathbf{x}_3^*) = \mu_{\mathbf{B}}(\mathbf{x}_3^*) = \mu_{\mathbf{B}}(\mathbf{x}) \wedge \mu_{\mathbf{B}}(\mathbf{x}) \text{ transveral points} \end{array}\right\} \text{Actual-Potential Polarit} \tag{1}$$

Similarly, the analysis of actual-potential polarity may be represented in a dualistic form of positive (actual) pole and negative (potential) pole of variety transformation in terms of existence and nonexistence dualities. Every pole of the polarity has existence and nonexistence characteristic subsets in such framework that in existence is nonexistence and in nonexistence is existence in dominating structures. The simultaneity of existence and nonexistence in each pole is a powerful methodological integration of the fuzzy paradigm and the principle of opposites that offers an approach to deal with qualitative-quantitative process of transformation under any subjective-objective phenomena over the epistemological space. Like the triangular fuzzy number (TFN), one may choose a continuous functional representation of the membership set. These functions include exponential, s-shape, ramp and many others depending on the nature of the transformation decision-choice problem. An example of the use of continuous characteristic function is show as an epistemic geometry in Fig. 3.

Mathematically, we may specify the dualistic representation of the existence-nonexistence polarity of a variety transformation process in terms of variety's respective poles as

$$\left.\begin{array}{l} \mathbf{A} = \{(\mathbf{x}, \mu_{\mathbf{A}}(\mathbf{x}))|x \in \mathbf{E}, \mu_{\mathbf{A}}(\mathbf{x}) \in [0,1] \text{ and } (d\mu/dx) \geq 0\} \text{Potential Pole} \\ \mathbf{B} = \{(\mathbf{x}, \mu_{\mathbf{B}}(\mathbf{x}))|x \in \mathbf{E}, \mu_{\mathbf{B}}(\mathbf{x}) \in [0,1] \text{ and } (d\mu/dx) \leq 0\} \text{Actual Pole} \end{array}\right\} \text{Existence-Nonexistence Polarity} \tag{2}$$

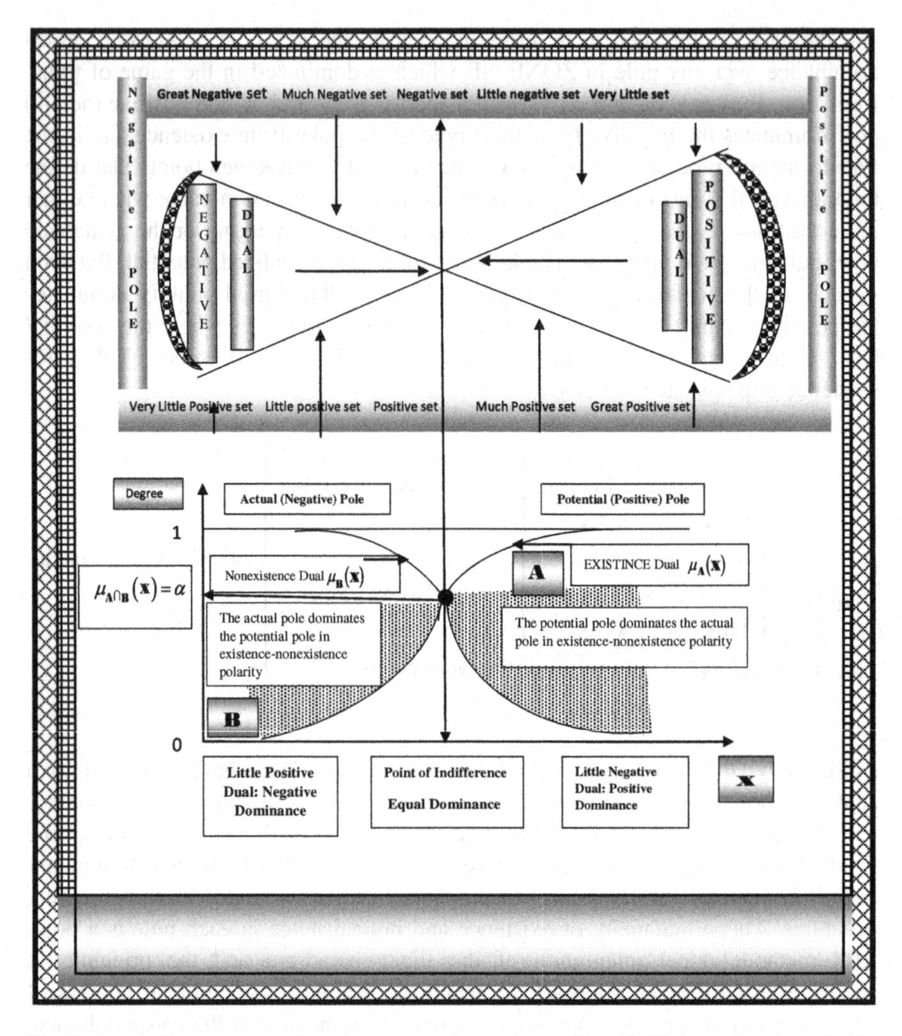

Fig. 3 A cognitive geometry of representation of negative-positive duality in continuum and unity in relation to the fuzzy paradigm, approximate reasoning in variety-transformation for info-dynamics

Given this representation we may then define fuzzy game as a transformation decision problem as $\Delta = \{\mathbf{A} \cap \mathbf{B} | \mathbf{A}, \mathbf{B} \subset \mathbf{E}\}$, whose membership characteristic function may be defined from the combination of the membership characteristic functions of existence and nonexistence as in Eq. 2, where $\mathbf{E}$ is the space of variety existence as:

$$\Delta = \{(\mathbf{x}, \mu_{\Delta}(\mathbf{x}))|\mu_{\Delta}(\mathbf{x}) = \mu_{\mathbf{A}}(\mathbf{x}) \wedge \mu_{\mathbf{A}}(\mathbf{x}) \text{ and } \mathbf{x} \in \mathbf{E}\} \tag{3}$$

The variety-transformation decision problem is to find the transversality point that may be used for fuzzy decomposition into existence and nonexistence. The solution is obtained from the defining the fuzzy optimizing the membership decision function of Eq. 3

$$\operatorname*{opt}_{\mathbf{x}\in\mathbf{E}} \mu_{\Delta}(\mathbf{x}) = \begin{cases} \operatorname*{opt}_{\mathbf{x}\in\mathbf{Q}} \mu_{\mathbf{A}}(\mathbf{x}) \\ \text{s.t } \mathbf{Q} = \{\mathbf{x} \in \mathbf{E}|[\mu_{\mathbf{A}}(\mathbf{x}) - \mu_{\mathbf{B}}(\mathbf{x})] \geq 0\} \end{cases} \tag{4}$$

The method for abstracting the solution may be found [R5.13, R5.41, R6, R6.3, R6.6, R6.8].The solution to the fuzzy optimization to variety transformation problem may be characterized as:

$$\operatorname{Opt} \mu_{\Delta}(\mathbf{x}) = \{(\mathbf{x}^*, \mu_{\Delta}(\mathbf{x}^*))|\mu_{\mathbf{A}}(\mathbf{x}^*) = \mu_{\mathbf{B}}(x^*) = \mu_{\Delta}(\mathbf{x}^*) = \alpha \in [0, 1], \mathbf{x}^* \in \mathbf{E}\} \tag{5}$$

The value $\alpha^* \in [0, 1]$ of Eq. 5 is the transversality point between the variety's existence and the variety's nonexistence. This value is used for establishing the needed fuzzy decomposition of existence and nonexistence in the form which is used

$$\left.\begin{array}{l} \mathbf{E} = \{(\mathbf{x}, \mu_{\mathbb{A}})|\mathbf{x} > \mathbf{x}^* \text{ and } \mu_{\mathbb{A}}(\mathbf{x}^*) > \alpha^*, \forall \mathbf{x} \in \mathbf{x}\} \text{ Existence Pole} \\ \mathbf{NE} = \{(\mathbf{x}, \mu_{\mathbb{A}})|\mathbf{x} \leq \mathbf{x}^* \text{ and } \mu_{\mathbb{A}}(\mathbf{x}^*) \leq \alpha^*, \forall \mathbf{x} \in \mathbf{E}\} \text{ Nonexistence Pole} \\ \mu_{\mathbb{A}}(\mathbf{x}^*) = \alpha^* \text{is the transversality point} \end{array}\right\} \text{Actual-Potential Polarity} \tag{6}$$

The extensively available functional forms of characteristic membership functions create an important flexibility in dealing with all dualities residing in the poles of polarities as well as supporting the indicator functions that are decision-choice-based in the construct of possibility and probability sets in the analysis of information, uncertainties of transformational events in the development of the analytical framework in understanding internal transformation of varieties and production of info-flows. The added advantage is that it provides a framework in the development of the theory of general entropy from the fuzzy domain. The development of the general entropy must be seen from the framework of epistemological information as the third dimension of existence, and time as the fourth dimension of existence in understanding uncertainty as a dynamic phenomenon where variety transformation is time dependent. Such a fuzzy framework and the development of corresponding entropy is a suggested direction in future research. Here fuzziness must be seen as a foundational paradigm and not as a methodological grafting through a simple method of fuzzification of the classical information variables which must be distinguished from fuzzy information variables.

Multidisciplinary References

R1. Category Theory in Mathematics, Logic and Sciences

[R1.1] Awodey, S., "Structure in Mathematics and Logic: A Categorical Perspective," *Philosophia Mathematica*, Vol. 3. 1996, pp. 209–237.

[R1.2] Bell, J. L., "Category Theory and the Foundations of Mathematics," *British Journal of Science*, Vol. 32, 1981, pp. 349–358.

[R1.3] Bell, J. L., "Categories, Toposes and Sets," *Synthese*, Vol. 51, 1982, pp. 393–337.

[R1.4] Black, M., *The Nature of Mathematics*, Totowa, N.J., Littlefield, Adams and Co., 1965.

[R1.5] Blass, A., "The Interaction between Category and Set Theory," *Mathematical Applications of Category Theory*, Vol. 30, 1984, pp. 5–29.

[R1.6] Brown, B. and J Woods (eds.), *Logical Consequence; Rival Approaches and New Studies in exact Philosophy: Logic, Mathematics and Science*, Vol. II Oxford, Hermes, 2000.

[R1.7] Domany, J. L., et al., *Models of Neural Networks III*, New York Springer, 1996.

[R1.8] Feferman, S., "Categorical Foundations and Foundations of Category Theory," in R. Butts (ed.), *Logic, Foundations of Mathematics and Computability*, Boston, Mass., Reidel, 1977, pp. 149–169.

[R1.9] Glimcher, P. W., *Decisions, Uncertainty, and the Brain: The Science of Neoroeconomics*, Cambridge, Mass., MIT Press, 2004.

[R1.10] Gray, J. W. (ed.) *Mathematical Applications of Category Theory* (American Mathematical Society Meeting 89th Denver Colo. 1983), Providence, R.I., American Mathematical Society, 1984.

[R.1.11] Johansson, Ingvar, *Ontological Investigations: An Inquiry into the Categories of Nature, Man, and Society*, New York, Routledge, 1989.

[R1.12] Kamps, K. H., D. Pumplun, and W. Tholen (eds.) *Category Theory: Proceedings of the International Conference*, Gummersbach, July 6–10, New York, Springer, 1982.

[R1.13] Landry, E., Category Theory: the Language of Mathematics," *Philosophy of Science*, Vol. 66, (Supplement), S14–S27.

[R1.14] Landry E. and J.P Marquis, "Categories in Context: Historical, Foundational and Philosophical," *Philiosophia Mathematica*, Vol. 13, 2005, pp. 1–43.

[R1.15] Marquis, J.-P., "Three Kinds of Universals in Mathematics," in B. Brown, and J. Woods (eds.), *Logical Consequence; Rival Approaches and New Studies in exact Philosophy: Logic, Mathematics and Science*, Vol. II Oxford, Hermes, 2000, pp. 191–212.

[R1.16] McLarty, C., "Category Theory in Real Time," *Philosophia Mathematica*, Vol. 2, 1994, pp. 36–44.

[R1.17] McLarty, C., "Learning from Questions on Categorical Foundations," *Philosophia Mathematica*, Vol. 13, 2005, pp. 44–60.

[R1.18] Ross, Don, *Economic theory and Cognitive Science; Microexplanation*, Cambridge, Mass., MIT Press, 2005.

K.K. Dompere, *The Theory of Info-Dynamics: Rational Foundations of Information-Knowledge Dynamics*, Studies in Systems, Decision and Control 114, https://doi.org/10.1007/978-3-319-63853-9

[R1.19] Rodabaugh, S. et al., (eds.), *Application of Category Theory to Fuzzy Subsets*, Boston, Mass., Kluwer, 1992.
[R1.20] Sieradski, Allan, J., An Introduction to Topology and Homotopy, PWS-KENT Pub. Boston, 1992.
[R1.20] Taylor, J. G. (ed.), *Mathematical Approaches to Neural Networks*, New York, North-Holland, 1993.
[R1.21] Van Benthem, J. et al. (eds.), *The Age of Alternative Logics: Assessing Philosophy of Logic and Mathematics Today*, New York, Springer, 2006.

R2. Concepts of Information, Fuzzy Probability, Fuzzy Random Variable and Random Fuzzy Variable

[R2.1] Bandemer, H., "From Fuzzy Data to Functional Relations," *Mathematical Modelling*, Vol. 6, 1987, pp. 419–426.
[R2.2] Bandemer, H. et al., *Fuzzy Data Analysis*, Boston, Mass, Kluwer, 1992.
[R2.3] Kruse, R. et al., *Statistics with Vague Data*, Dordrecht, D. Reidel Pub. Co., 1987.
[R2.4] El Rayes, A. B. et al., "Generalized Possibility Measures," *Information Sciences*, Vol. 79, 1994, pp. 201–222.
[R2.5] Dumitrescu, D., "Entropy of a Fuzzy Process," *Fuzzy Sets and Systems*, Vol. 55, #2, 1993, pp. 169–177.
[R2.6] Delgado, M. et al., "On the Concept of Possibility-Probability Consistency," *Fuzzy Sets and Systems*, Vol. 21, #3, 1987, pp. 311–318.
[R2.7] Devi, B. B. et al., "Estimation of Fuzzy Memberships from Histograms," *Information Sciences*, Vol. 35, #1, 1985, pp. 43–59.
[R2.8] Dubois, D. et al., "Fuzzy Sets, Probability and Measurement," *European Jour. of Operational Research*, Vol. 40, #2, 1989, pp. 135–154.
[R2.9] Fruhwirth-Schnatter, S., "On Statistical Inference for Fuzzy Data with Applications to Descriptive Statistics," *Fuzzy Sets and Systems*, Vol. 50, #2, 1992, pp. 143–165.
[R2.10] Gaines, B. R., "Fuzzy and Probability Uncertainty logics," *Information and Control*, Vol. 38, #2, 1978, pp. 154–169.
[R2.11] Geer, J. F. et al., "Discord in Possibility Theory," *International Jour. of General Systems*, Vol. 19, 1991, pp. 119–132.
[R2.12] Geer, J. F. et al., "A Mathematical Analysis of Information-Processing Transformation Between Probabilistic and Possibilistic Formulation of Uncertainty," *International Jour. of General Systems*, Vol. 20, #2, 1992, pp. 14–176.
[R2.13] Goodman, I. R. et al., *Uncertainty Models for Knowledge Based Systems*, New York, North-Holland, 1985.
[R2.14] Grabish, M. et al., *Fundamentals of Uncertainty Calculi with Application to Fuzzy Systems*, Boston, Mass., Kluwer, 1994.
[R2.15] Guan, J. W. et al., *Evidence Theory and Its Applications*, Vol. 1, New York, North-Holland, 1991.
[R2.16] Guan, J. W. et al., *Evidence Theory and Its Applications*, Vol. 2, New York, North-Holland, 1992.
[R2.17] Hisdal, E., Are Grades of Membership Probabilities?," *Fuzzy Sets and Systems*, Vol. 25, #3, 1988, pp. 349–356.
[R2.18] Höhle Ulrich, "A Mathematical Theory of Uncertainty," in R.R. Yager (ed.) *Fuzzy Set and Possibility Theory: Recent Developments*, New York, Pergamon, 1982, pp. 344–355.
[R2.19] Kacprzyk, Janusz and Mario Fedrizzi (eds.) *Combining Fuzzy Imprecision with Probabilistic Uncertainty in Decision Making*, New York, Plenum Press, 1992.

[R2.20] Kacprzyk, J. et al., *Combining Fuzzy Imprecision with Probabilistic Uncertainty in Decision Making*, New York, Springer, 1988.

[R2.21] Klir, G. J., "Where Do we Stand on Measures of Uncertainty, Ambiguity, Fuzziness and the like?" *Fuzzy Sets and Systems*, Vol. 24, #2, 1987, pp. 141–160.

[R2.22] Klir, G. J. et al., *Fuzzy Sets, Uncertainty and Information*, Englewood Cliff, Prentice Hall, 1988.

[R2.23] Klir, G. J. et al., "Probability-Possibility Transformations: A Comparison," *Intern. Jour. of General Systems*, Vol. 21, #3, 1992, pp. 291–310.

[R2.24] Kosko, B., "Fuzziness vs Probability," *Intern. Jour. of General Systems*, Vol. 17, #(1–3) 1990, pp. 211–240.

[R2.26] Manton, K. G. et al., *Statistical Applications Using Fuzzy Sets*, New York, Wiley, 1994.

[R2.27] Meier, W., et al., "Fuzzy Data Analysis: Methods and Industrial Applications," *Fuzzy Sets and Systems*, Vol. 61, #1, 1994, pp. 19–28.

[R2.28] Nakamura, A., et al., "A logic for Fuzzy Data Analysis," *Fuzzy Sets and Systems*, Vol. 39, #2, 1991, pp. 127–132.

[R2.29] Negoita, C. V. et al., *Simulation, Knowledge-Based Compting and Fuzzy Statistics*, New York, Van Nostrand Reinhold, 1987.

[R2.30] Nguyen, H. T., "Random Sets and Belief Functions," *Jour. of Math. Analysis and Applications*, Vol. 65, #3, 1978, pp. 531–542.

[R2.31] Prade, H. et al., "Representation and Combination of Uncertainty with belief Functions and Possibility Measures," *Comput. Intell.*, Vol. 4, 1988, pp. 244–264.

[R2.32] Puri, M. L. et al., "Fuzzy Random Variables," *Jour. of Mathematical Analysis and Applications*, Vol. 114, #2, 1986, pp. 409–422.

[R2.33] Rao, N. B. and A. Rashed, "Some Comments on Fuzzy Random Variables," *Fuzzy Sets and Systems*, Vol. 6, # 3, 1981, pp. 285–292.

[R2.34] Sakawa, M. et al., "Multiobjective Fuzzy linear Regression Analysis for Fuzzy Input-Output Data," *Fuzzy Sets and Systems*, Vol. 47, #2, 1992, pp. 173–182.

[R2.35] Schneider, M. et al., "Properties of the Fuzzy Expected Values and the Fuzzy Expected Interval," *Fuzzy Sets and Systems*, Vol. 26, #3, 1988, pp. 373–385.

[R2.36] Stein, N. E. and K. Talaki, "Convex Fuzzy Random Variables," *Fuzzy Sets and Systems*, Vol. 6, #3, 1981, pp. 271–284.

[R2.37] Sudkamp, T., "On Probability-Possibility Transformations," *Fuzzy Sets and Systems*, Vol. 51, #1, 1992, pp. 73–82.

[R2.38] Walley, P., *Statistical Reasoning with Imprecise Probabilities*, London Chapman and Hall, 1991.

[R2.39] Wang, G. Y. et al., "The Theory of Fuzzy Stochastic Processes," *Fuzzy Sets and Systems*, Vol. 51, #2 1992, pp. 161–178.

[R2.40] Zadeh, L. A., "Probability Measure of Fuzzy Event," *Jour. of Math Analysis and Applications*, Vol. 23, 1968, pp. 421–427.

R3. Exact Science, Inexact Sciences and Information

[R3.1] Achinstein, P., "The Problem of Theoretical Terms," in Brody, Baruch A. (Ed.) *Reading in the Philosophy of Science*, Englewood Cliffs, NJ, Prentice Hall, 1970.

[R3.2] Amo Afer, A. G., *The Absence of Sensation and the Faculty of Sense in the Human Mind and Their Presence in our Organic and Living Body, Dissertation and Other essays 1727-1749,* Halle Wittenberg, Jena, Martin Luther University Translation, 1968.

[R3.3] Beeson, M. J., *Foundations of Constructive Mathematics*, Berlin/New York, Springer, 1985.

[R3.4] Benacerraf, P., "God, the Devil and Gödel," *Monist,* Vol. 51, 1967, pp. 9–32.

[R3.5] Benecerraf, P and H. Putnam (eds.), *Philosophy of Mathematics: Selected Readings*, Cambridge, Cambridge University Press, 1983.

[R3.6] Black, Max, *The Nature of Mathematics,* Totowa, Littlefield, Adams and Co. 1965.
[R3.7] Blanche, R., *Contemporary Science and Rationalism*, Edinburgh, Oliver and Boyd, 1968.
[R3.8] Blanshard, Brand, *The Nature of Thought*, London, Allen and Unwin, 1939.
[R3.9] Blauberg, I. V., V. N. Sadovsky and E. G. Yudin, Systems Theory: Philosophical and Methodological Problems, Moscow, Progress Publishers, 1977.
[R3.10] Braithwaite, R. B., *Scientific Explanation*, Cambridge, Cambridge University Press, 1955.
[R3.11] Brody, Baruch A. (ed.), *Reading in the Philosophy of Science*, Englewood Cliffs, N.J., Prentice Hall, 1970.
[R3.12] Brody, Baruch A., "Confirmation and Explanation," in Brody, Baruch A. (ed.) *Reading in the Philosophy of Science*, Englewood Cliffs, N.J., Prentice-Hall, 1970, pp. 410–426.
[R3.13] Brouwer, L. E. J., "Intuitionism and Formalism", *Bull of American Math. Soc.*, Vol. 20, 1913, pp. 81–96.; Also in Benecerraf, P. and H. Putnam (eds.), *Philosophy of Mathematics: Selected Readings*, Cambridge, Cambridge University Press, 1983. pp. 77–89.
[R3.14] Brouwer, L. E. J., "Consciousness, Philosophy, and Mathematics," in Benecerraf, P. and H. Putnam (eds.), *Philosophy of Mathematics: Selected Readings*, Cambridge, Cambridge University Press, 1983, pp. 90–96.
[R3.15] Brouwer, L. E. J., *Collected Works, Vol. 1: Philosophy and Foundations of Mathematics* [A Heyting (ed.)], New York, Elsevier, 1975.
[R3.16] Campbell, Norman R., *What is Science?*, New York, Dover, 1952.
[R3.17] Carnap, R., "Foundations of Logic and Mathematics," in *International Encyclopedia of Unified Science*, Chicago, Univ. of Chicago, 1939, pp. 143–211.
[R3.18] Carnap, Rudolf, "On Inductive Logic," *Philosophy of Science*, Vol. 12, 1945, pp. 72–97.
[R3.19] Carnap, Rudolf, "The Methodological Character of Theoretical Concepts," in Herbert Feigl and M. Scriven (eds.) *Minnesota Studies in the Philosophy of Science, Vol. I*, 1956, pp. 38–76.
[R3.20] Charles, David and Kathleen Lennon (eds.), *Reduction, Explanation, and Realism*. Oxford, Oxford University Press, 1992.
[R3.21] Cohen, Robert S. and Marx W. Wartofsky (eds.), *Methodological and Historical Essays in the Natural and Social Sciences*, Dordrecht, D. Reidel Publishing Co. 1974.
[R3.22] Dalen van, D. (ed.), *Brouwer's Cambridge Lectures on Intuitionism,* Cambridge, Cambridge University Press, 1981.
[R3.23] Davidson, Donald, *Truth and Meaning: Inquiries into Truth and Interpretation*, Oxford, Oxford University Press, 1984.
[R3.24] Davis, M., *Computability and Unsolvability*, New York, McGraw-Hill, 1958.
[R3.25] Denonn. Lester E. (ed.), *The Wit and Wisdom of Bertrand Russell*, Boston, MA., The Beacon Press, 1951.
[R3.26] Dummett, M., "The Philosophical Basis of Intuitionistic Logic," in Benecerraf, P. and H. Putnam (eds.), *Philosophy of Mathematics: Selected Readings*, Cambridge, Cambridge University Press, 1983, pp. 97–129.
[R3.27] Feigl, Herbert and M. Scriven (eds.), *Minnesota Studies in the Philosophy of Science,* Vol. I, 1956.
[R3.28] Feigl, Herbert and M. Scriven (eds.), *Minnesota Studies in the Philosophy of Science*, Vol. II, 1958.
[R3.29] Garfinkel, Alan, *Forms of Explanation: Structures of Inquiry in Social Science,* New Haven, Conn., Yale University Press, 1981.
[R3.30] George, F. H., *Philosophical Foundations of Cybernetics*, Tunbridge Well, Great Britain, 1979.
[R3.31] Gillam, B., "Geometrical Illusions," *Scientific American*, January, 1980, pp. 102–111.
[R3.32] Gödel, Kurt., "What is Cantor's Continuum Problem?" in Benecerraf, P. and H. Putnam (eds.), *Philosophy of Mathematics: Selected Readings*, Cambridge, Cambridge University Press, 1983. pp. 470–486.
[R3.33] Gorsky, D. R., *Definition,* Moscow, Progress Publishers, 1974.

[R3.34] Gray, William and Nicholas D. Rizzo (eds.), *Unity Through Diversity*. New York, Gordon and Breach, 1973.
[R3.35] Hart, W. D. (ed.), *The Philosophy of Mathematics*, Oxford, Oxford University Press, 1996.
[R3.36] Hartkamper, A and H. Schmidt, Structure and Approximation in Physical Theories, New York, Plenum Press, 1981.
[R3.37] Hausman, David, M., *The Exact and Separate Science of Economics,* Cambridge, Cambridge University Press, 1992.
[R3.38] Helmer, Olaf and Nicholar Rescher, *On the Epistemology of the Inexact Sciences*, P-1513, Santa Monica, CA, Rand Corporation, October 13, 1958.
[R3.39] Hempel, C. G., "Studies in the Logic of Confirmation," *Mind*, Vol. 54, Part I, 1945, pp. 1–26.
[R3.34] Hempel, Carl G., "The Theoretician's Dilemma," in Herbert Feigl and M. Scriven (eds.) *Minnesota Studies in the Philosophy of Science*, Vol. II, 1958, pp. 37–98.
[R3.35] Hempel, C. G. and P. Oppenheim, "Studies in the Logic of Explanation," *Philosophy of Science*, Vol. 15, 1948, pp. 135–175. [also in Brody, Baruch A. (ed.) *Reading in the Philosophy of Science*, Englewood Cliffs, NJ., Prentice-Hall, 1970, pp. 8–27.
[R3.36] Heyting, A., *Intuitionism: An Introduction,* Amsterdam: North-Holland, 1971.
[R3.37] Hintikka, Jackko (ed.), *The Philosophy of Mathematics*, London, Oxford University Press, 1969.
[R3.38] Hockney D. et al. (eds.), *Contemporary Research in Philosophical Logic and Linguistic Semantics*, Dordrecht-Holland, Reidel Pub., Co. 1975.
[R3.39] Hoyninggen-Huene, Paul and F. M. Wuketits, (eds.), *Reductionism and Systems Theory in the Life Science: Some Problems and Perspectives*, Dordrecht, Kluwer Academic Pub. 1989.
[R3.40] Ilyenkov, E. V., *Dialectical Logic: Essays on Its History and Theory*, Moscow, Progress Publishers, 1977.
[R3.41] Kedrov, B. M., "Toward the Methodological Analysis of Scientific Discovery," *Soviet Studies in Philosophy*, Vol. 11962, pp. 45–65.
[R3.42] Kemeny, John G, and P Oppenheim, "On Reduction," in Brody, Baruch A. (ed.) *Reading in the Philosophy of Science*, Englewood Cliffs, NJ., Prentice-Hall, 1970, pp. 307–318.
[R3.43] Klappholz, K., "Value Judgments of Economics," *British Jour. of Philosophy*, Vol. 15, 1964, pp. 97–114.
[R3.44] Kleene, S. C., "On the Interpretation of Intuitionistic Number Theory," Journal of Symbolic Logic, Vol. 10, 1945, pp. 109–124.
[R3.45] Kmita, Jerzy, "The Methodology of Science as a Theoretical Discipline," *Soviet Studies in Philosophy*, Spring, 1974, pp. 38–49.
[R3.46] Krupp, Sherman R., (ed.), *The Structure of Economic Science*, Englewood Cliff, N.J., Prentice-Hall, 1966.
[R3.47] Kuhn, T., *The Structure of Scientific Revolution,* Chicago, University of Chicago Press, 1970.
[R3.48] Kuhn, Thomas, "The Function of Dogma in Scientific Research," in Brody, Baruch A. (ed.) *Reading in the Philosophy of Science*, Englewood Cliffs, NJ., Prentice-Hall, 1970 pp. 356–374.
[R3.49] Kuhn, Thomas, *The Essential Tension: Selected Studies in Scientific Tradition and Change,* Chicago, University of Chicago Press, 1979.
[R3.50] Lakatos, I. (ed.), *The Problem of Inductive Logic,* Amsterdam, North Holland, 1968.
[R3.51] Lakatos, I., *Proofs and Refutations: The Logic of Mathematical Discovery*, Cambridge, Cambridge University Press, 1976.
[R3.52] Lakatos, I., *Mathematics, Science and Epistemology: Philosophical Papers Vol. 2*, edited by J. Worrall and G. Currie, Cambridge, Cambridge Univ. Press, 1978.
[R3.53] Lakatos, I., *The Methodology of Scientific Research Programmes*, Vol. 1, New York, Cambridge University Press, 1978.

[R3.54] Lakatos, Imre and A. Musgrave (eds.), *Criticism and the Growth of Knowledge*, New York, Cambridge University Press, 1979. Holland, 1979, pp. 153–164.

[R3.55] Lawson, Tony, *Economics and Reality*, New York, Routledge, 1977.

[R3.56] Lenzen, Victor, "Procedures of Empirical Science," in Neurath, Otto et al. (eds.), *International Encyclopedia of Unified Science,* Vol. 1–10, Chicago, University of Chicago Press, 1955, pp. 280–338.

[R3.57] Levi, Isaac, "Must the Scientist make Value Judgments?," in Brody, Baruch A. (Ed.) *Reading in the Philosophy of Science*, Englewood Cliffs, NJ., Prentice-Hall, 1970, pp. 559–570.

[R3.58] Tse-tung, *Mao, On Practice and Contradiction*, in *Selected works of Mao Tse-tung*, Piking, 1937. Also, London, Revolutions, 2008.

[R3.58] Lewis, David, *Convention: A Philosophical Study,* Cambridge, Mass., Harvard University Press, 1969.

[R3.59] Mayer, Thomas, *Truth versus Precision in Economics*, London, Edward Elgar, 1993.

[R3.60] Menger, Carl, *Investigations into the Method of the Social Sciences with Special Reference to Economics*, New York, New York University Press, 1985.

[R3.61] Mirowski, Philip (ed.), *The Reconstruction of Economic Theory*, Boston, Mass. Kluwer Nijhoff, 1986.

[R3.62] Mueller, Ian, *Philosophy of Mathematics and Deductive Structure in Euclid's Elements*, Cambridge, Mass., MIT Press, 1981.

[R3.63] Nagel, Ernest, "Review: Karl Niebyl, Modern Mathematics and Some Problems of Quantity, Quality, and Motion in Economic Analysis," *The Journal of Symbolic Logic*, 1940, p. 74.

[R3.64] Nagel, E. et al. (ed.), *Logic, Methodology, and the Philosophy of Science*, Stanford, Stanford University Press, 1962.

[R3.65] Narens, Louis, "A Theory of Belief for Scientific Refutations," *Synthese*, Vol. 145, 2005, pp. 397–423.

[R3.66] Narskii, I. S., "On the Problem of Contradiction in Dialectical Logic," *Soviet Studies in Philosophy*, Vol. vi, #4 pp. 3–10, 1965.

[R3.67] Neurath, Otto et al. (eds.), *International Encyclopedia of Unified Science,* Vol. 1–10, Chicago, University of Chicago Press, 1955.

[R3.68] Neurath Otto, "Unified Science as Encyclopedic," in Neurath, Otto et al. (eds.), *International Encyclopedia of Unified Science*, Vol. 1–10, Chicago, University of Chicago Press, 1955, pp. 1–27.

[R3.69] Planck, Max, *Scientific Autobiography and Other Papers*, Westport, Conn. Greenwood, 1971.

[R3.70] Planck, Max, "The Meaning and Limits of Exact Science," in Max Planck, *Scientific Autobiography and Other Papers,* Westport, Conn. Greenwood, 1971, pp. 80–120.

[R3.71] Polanyi, Michael, "Genius in Science," in Robert S. Cohen, and Marx W. Wartofsky (eds.), *Methodological and Historical Essays in the Natural and Social Sciences*, Dordrecht, D. Reidel Publishing Co. 1974, pp. 57–71.

[R3.72] Popper, Karl, *The Nature of Scientific Discovery*, New York, Harper and Row, 1968.

[R3.73] Putnam, Hilary., "Models and Reality," in Benecerraf, P. and H. Putnam (eds.), *Philosophy of Mathematics: Selected Readings*, Cambridge, Cambridge University Press, 1983. pp. 421–444.

[R3.74] Reise, S., *The Universe of Meaning*, New York, The Philosophical Library, 1953.

[R3.75] Robinson, R., *Definition*, Oxford, Clarendon Press, 1950.

[R3.76] Rudner, Richard, "The Scientist qua Scientist Makes Value Judgments," *Philosophy of Science*, Vol. 20, 1953, pp. 1–6.

[R3.77] Russell, B., *Our Knowledge of the External World*, New York, Norton, 1929.

[R3.78] Russell, B., *Human Knowledge, Its Scope and Limits*, London, Allen and Unwin, 1948.

[R3.79] Russell, B., *Logic and Knowledge: Essays 1901-1950*, New York, Capricorn Books, 1971.

[R3.80] Russell, B., *An Inquiry into Meaning and Truth*, New York, Norton, 1940.
[R3.81] Russell, Bertrand, *Introduction to Mathematical Philosophy*, London, George Allen and Unwin, 1919.
[R3.82] Russell, Bertrand, *The Problems of Philosophy*, Oxford, Oxford University Press, 1978.
[R3.83] Rutkevih, M. N., "Evolution, Progress, and the Law of Dialectics," *Soviet Studies in Philosophy,* Vol. IV, #3, pp. 34–43, 1965.
[R3.84] Ruzavin, G. I., "On the Problem of the Interrelations of Modern Formal Logic and Mathematical Logic," *Soviet Studies in Philosophy*, Vol. 3, #1, 1964, pp. 34–44.
[R3.85] Scriven, Michael, "Explanations, Predictions, and Laws," in Brody, Baruch A. (ed.) *Reading in the Philosophy of Science, Englewood Cliffs*, NJ., Prentice-Hall, 1970, pp. 88–104.
[R3.86] Sellars, Wilfrid, The Language of Theories," in Brody, Baruch A. (ed.) *Reading in the Philosophy of Science,* Englewood Cliffs, NJ., Prentice-Hall, 1970, pp. 343–353.
[R3.89] Sterman, John, "The Growth of Knowledge: Testing a Theory of Scientific Revolutions with a Formal Model," *Technological Forecasting and Social Change*, Vol. 28, 1995, pp. 93–122.
[R3.90] Tsereteli, S. B. "On the Concept of Dialectical Logic,", *Soviet Studies in Philosophy*, Vol. V, #2, pp. 15–21, 1966.
[R3.91] Tullock, Gordon, *The Organization of Inquiry,* Indianapolis, Indiana, Liberty Fund Inc., 1966.
[R3.92] Van Fraassen, B., *Introduction to Philosophy of Space and Time*, New York, Random House, 1970.
[R3.93] Veldman, W., "A Survey of Intuitionistic Descriptive Set Theory," in P.P. Petkov (ed.), Mathematical Logic: Proceedings of the Heyting Conference, New York, Plenum Press, 1990, pp. 155–174.
[R3.94] Vetrov, A. A., "Mathematical Logic and Modern Formal Logic," *Soviet Studies in Philosophy*, Vol. 3, #1, 1964, pp. 24–33.
[R3.95] von Mises, Ludwig, *Epistemological Problems in Economics*, New York, New York University Press, 1981.
[R3.96] Wang, Hao, *Reflections on Kurt* Gödel, Cambridge, Mass. MIT Press, 1987.
[R3.97] Watkins, J. W. N., "The Paradoxes of Confirmation", in Brody, Baruch A. (ed.) *Reading in the Philosophy of Science*, Englewood Cliffs, NJ., Prentice-Hall, 1970, pp. 433–438.
[R3.98] Whitehead, Alfred North, *Process and Reality*, New York, The Free Press, 1978.
[R3.99] Wittgenstein, Ludwig, *Ttactatus Logico-philosophicus,* Atlantic Highlands, N.J., The Humanities Press Inc., 1974.
[R3.100] Woodger, J. H., *The Axiomatic Method in Biology*, Cambridge, Cambridge University Press, 1937.
[R3.101] Zeman, Jiří, "Information, Knowledge and Time," in Libor Kubát, and J. Zeman (eds.), *Entropy and Information in the Physical Sciences*, Amsterdam, Elsevier, 1975, pp. 245–260.

R4. Fuzzy Logic, Information and Knowledge-Production

[R4.1] Baldwin, J. F., "A New Approach to Approximate Reasoning Using a Fuzzy Logic," *Fuzzy Sets and Systems*, Vol. 2, #4, 1979, pp. 309–325.
[R4.2] Baldwin, J. F., "Fuzzy Logic and Fuzzy Reasoning," *Intern. J. Man-Machine Stud.*, Vol. 11, 1979, pp. 465–480.
[R4.3] Baldwin, J. F., "Fuzzy Logic and Its Application to Fuzzy Reasoning," in. M. M. Gupta et al. (eds.), *Advances in Fuzzy Set Theory and Applications*, New York, North-Holland, 1979, pp. 96–115.
[R4.4] Baldwin, J. F. et al., "Fuzzy Relational Inference Language," *Fuzzy Sets and Systems*, Vol. 14, #2, 1984, pp. 155–174.
[R4.5] Baldsin, J. and B. W. Pilsworth, "Axiomatic Approach to Implication For Approximate Reasoning With Fuzzy Logic," *Fuzzy Sets and Systems*, Vol. 3, #2, 1980, pp. 193–219.

[R4.6] Baldwin, J. F. et al., "The Resolution of Two Paradoxes by Approximate Reasoning Using A Fuzzy Logic," *Synthese*, Vol. 44, 1980, pp. 397–420.

[R4.7] Dompere, K. K., *Fuzzy Rationality: Methodological Critique and Unity of Classical, Bounded and Other Rationalities*, (Studies in Fuzziness and Soft Computing, vol. 235) New York, Springer, 2009.

[R4.8] Dompere Kofi K., *Epistemic Foundations of Fuzziness*, (Studies in Fuzziness and Soft Computing, vol. 236) New York, Springer, 2009.

[R4.9] Dompere Kofi K., *Fuzziness and Approximate Reasoning: Epistemics on Uncertainty, Expectation and Risk in Rational Behavior*, (Studies in Fuzziness and Soft Computing, vol. 237) New York, Springer, 2009.

[R4.10] Dompere, Kofi K., *The Theory of the Knowledge Square: The Fuzzy Rational Foundations of Knowledge-Production Systems*, New York, Springer, 2013.

[R4.11] Dompere, Kofi K., "Cost-Benefit Analysis, Benefit Accounting and Fuzzy Decisions: Part I, Theory", *Fuzzy Sets and Systems*, Vol. 92, 1997, pp. 275–287.

[R4.12] Dompere, Kofi K., "The Theory of Social Cost and Costing For Cost-Benefit Analysis in a Fuzzy Decision Space", Fuzzy Sets and Systems, Vol. 76, 1995, pp. 1–24.

[R4.13] Dompere, Kofi K., *Fuzzy Rational Foundations of Exact and Inexact Sciences*, New York, Springer, 2013.

[R4.14] Gaines, B. R., "Foundations of Fuzzy Reasoning," *Inter. Jour. of Man-Machine Studies*, Vol. 8, 1976, pp. 623–668.

[R4.15] Gaines, B. R., "Foundations of Fuzzy Reasoning," in Gupta, M.M. et al. (eds.), *Fuzzy Information and Decision Processes*, New York North-Holland, 1982, pp. 19–75.

[R4.16] Gaines, B. R., "Precise Past, Fuzzy Future," *International Journal. Of Man-Machine Studies.*, Vol. 19, #1, 1983, pp. 117–134.

[R4.17] Giles, R., "Lukasiewics Logic and Fuzzy Set Theory," *Intern. J. Man-Machine Stud.*, Vol. 8, 1976, pp. 313–327.

[R4.18] Giles, R., "Formal System for Fuzzy Reasoning," *Fuzzy Sets and Systems*, Vol. 2, #3, 1979, pp. 233–257.

[R4.19] Ginsberg, M. L. (ed.), *Readings in Non-monotonic Reason*, Los Altos, Ca., Morgan Kaufman, 1987.

[R4.20] Goguen, J. A., "The Logic of Inexact Concepts," *Synthese*, Vol. 19, 1969, pp. 325–373.

[R4.21] Gottinger, H. W., "Towards a Fuzzy Reasoning in the Behavioral Science," *Cybernetica*, Vol. 16, #2, 1973, pp. 113–135.

[R4.23] Gupta, M. M. et al., (eds.), *Approximate Reasoning In Decision Analysis*, North Holland, New York, 1982.

[R4.24] Höhle Ulrich, and E. P. Klement, *Non-Clasical Logics and their Applications to Fuzzy Subsets: A Handbook of the Mathematical Foundations of Fuzzy Set Theory*, Boston, Mass. Kluwer, 1995.

[R4.25] Kaipov, V., Kh. et al., "Classification in Fuzzy Environments," in M. M. Gupta et al. (eds.), *Advances in Fuzzy Set Theory and Applications,* New York, North-Holland, 1979, pp. 119–124,

[R4.26] Kaufman, A., "Progress in Modeling of Human Reasoning of Fuzzy Logic" in M. M. Gupta et al. (eds.), *Fuzzy Information and Decision Process*, New York, North-Holland, 1982, pp. 11–17.

[R4.27] Lakoff, G., "Hedges: A Study in Meaning Criteria and the Logic of Fuzzy Concepts," *Jour. Philos. Logic*, Vol. 2, 1973, pp. 458–508.

[R4.28] Lee, R. C. T., "Fuzzy Logic and the Resolution Principle," *Jour. of Assoc. Comput.* Mach., Vol. 19, 1972, pp. 109–119.

[R4.29] LeFaivre, R. A., "The Representation of Fuzzy Knowledge", *Jour. of Cybernetics*, Vol. 4, 1974, pp. 57–66.

[R4.30] Negoita, C. V., "Representation Theorems for Fuzzy Concepts," *Kybernetes*, Vol. 4, 1975, pp. 169–174.

[R4.31] Nowakowska, M., "Methodological Problems of Measurements of Fuzzy Concepts in Social Sciences", *Behavioral Sciences*, Vol. 22, #2, 1977, pp. 107–115.

[R4.33] Skala, H. J., "On Many-Valued Logics, Fuzzy Sets, Fuzzy Logics and Their Applications," *Fuzzy Sets and Systems*, Vol. 1, #2, 1978, pp. 129–149.

[R4.35] Van Fraassen, B. C., "Comments: Lakoff's Fuzzy Propositional Logic," in D. Hockney et al., (Eds.), *Contemporary Research in Philosophical Logic and Linguistic Semantics*, Holland, Reild, 1975, pp. 273–277.

[R4.36] Yager, R. R. et al. (eds.), An Introduction to Fuzzy Logic Applications in *Intelligent Systems*, Boston, Mass., Kluwer, 1992.

[R4.38] Zadeh, L. A., "Quantitative Fuzzy Semantics," *Inform. Science*, Vol. 3, 1971, pp. 159–176.

[R4.39] Zadeh, L. A., "A Fuzzy Set Interpretation of Linguistic Hedges," *Jour. Cybernetics*, Vol. 2, 1972, pp. 4–34.

[R4.41] Zadeh, L. A, "The Concept of a Linguistic Variable and Its Application to Approximate Reasoning," in K.S. Fu et al. (eds.), *Learning Systems and Intelligent Robots*, Plenum Press, New York, 1974, pp. 1–10.

[R4.42] Zadeh, L. A., et al., (eds.), *Fuzzy Sets and Their Applications to Cognitive and Decision Processes*, New York, Academic Press, 1974.

[R4.43] Zadeh, L. A., "The Birth and Evolution of Fuzzy Logic," *Intern. Jour. of General Systems*, Vol. 17, #(2–3) 1990, pp. 95–105.

R5. Fuzzy Mathematics and Paradigm of Approximate Reasoning Under Conditions of Inexactness and Vagueness

[R5.1] Bellman, R. E., "Mathematics and Human Sciences," in J. Wilkinson et al. (eds.), *The Dynamic Programming of Human Systems*, New York, MSS Information Corp., 1973, pp. 11–18.

[R5.2] Bellman, R. E and Glertz, M., "On the Analytic Formalism of the Theory of Fuzzy Sets," Information *Science*, Vol. 5, 1973, pp. 149–156.

[R5.3] Butnariu, D., Fixed Points for Fuzzy Mapping," *Fuzzy Sets and Systems*, Vol. 7, #2, pp. 191–207, 1982.

[R5.4] Butnariu, D., "Decompositions and Range for Additive Fuzzy Measures", *Fuzzy Sets and Systems*, Vol. 10, #2, pp. 135–155, 1983.

[R5.5] Chang, C. L., "Fuzzy Topological Spaces," *J. Math. Anal. and Applications*, Vol. 24, 1968, pp. 182–190.

[R5.6] Chang, S. S. L., "Fuzzy Mathematics, Man and His Environment", *IEEE Transactions on Systems, Man and Cybernetics*, SMC-2 1972, pp. 92–93,

[R5.7] Chang, S. S., "Fixed Point Theorems for Fuzzy Mappings," Fuzzy *Sets and Systems*, Vol. 17, 1985, pp. 181–187.

[R5.8] Chapin, E. W., "An Axiomatization of the Set Theory of Zadeh," *Notices, American Math. Society*, 687-02-4 754, 1971.

[R5.9] Chaudhury, A. K. and P. Das, "Some Results on Fuzzy Topology on Fuzzy Sets," *Fuzzy Sets and Systems,* Vol. 56, 1993, pp. 331–336.

[R5.10] Chitra, H., and P. V. Subrahmanyam, "Fuzzy Sets and Fixed Points," *Jour. of Mathematical Analysis and Application*, Vol. 124, 1987, pp. 584–590.

[R5.11] Czogala, J. et al., Fuzzy Relation Equations On a Finite Set," *Fuzzy Sets and Systems,* Vol. 7, #1, 1982. pp. 89–101.

[R5.12] DiNola, A. et al., (eds.), *The Mathematics of Fuzzy Systems*, Koln, Verlag TUV Rheinland, 1986.

[R5.13] Dompere, Kofi K., *Cost-Benefit Analysis and the Theory of Fuzzy Decisions: Identification and Measurement Theory* (Series: Studies in Fuzziness and Soft Computing, Vol. 158), Berlin, Heidelberg, Springer, 2004.

[R5.14] Dompere, Kofi K., *Cost-Benefit Analysis and the Theory of Fuzzy Decisions: Fuzzy Value Theory* (Series: Studies in Fuzziness and Soft Computing, Vol. 160), Berling, Heidelberg, Springer, 2004.

[R5.16] Dubois, D. and H. Prade, *Fuzzy Sets and Systems*, New York, Academic Press, 1980.

[R5.17] Dubois, "Fuzzy Real Algebra: Some Results," *Fuzzy Sets and Systems*, Vol. 2, #4, pp. 327–348, 1979.

[R5.18] Dubois, D. and H. Prade, "Gradual Inference Rules in Approximate Reasoning," *Information Sciences*, Vol. 61(1–2), 1992, pp. 103–122.

[R5.19] Dubois, D. and H. Prade, "On the Combination of Evidence in various Mathematical Frameworks." In: Flamm. J. and T. Luisi, (eds.), *Reliability Data Collection and Analysis*. Kluwer, Boston, 1992, pp. 213–241.

[R5.20] Dubois, D. and H. Prade, "Fuzzy Sets and Probability: Misunderstanding, Bridges and Gaps." *Proc. Second IEEE Intern. Conf. on Fuzzy Systems*, San Francisco, 1993, pp. 1059–1068.

[R5.21] Dubois, D. and H. Prade [1994], "A Survey of Belief Revision and Updating Rules in Various Uncertainty Models," *Intern. J. of Intelligent Systems*, Vol. 9, #1, pp. 61–100.

[R5.22] Filev, D. P. et al., "A Generalized Defuzzification Method via Bag Distributions," *Intern. Jour. of Intelligent Systems*, Vol. 6, #7, 1991, pp. 687–697.

[R5.23] Goetschel, R. Jr., et al., "Topological Properties of Fuzzy Number," *Fuzzy Sets and Systems*, Vol. 10, #1, pp. 87–99, 1983.

[R5.24] Goodman, I. R., "Fuzzy Sets As Equivalence Classes of Random Sets" in Yager, R.R. (ed.), *Fuzzy Set and Possibility Theory: Recent Development*, New York, Pergamon Press, 1992. pp. 327–343.

[R5.25] Gupta, M. M. et al., (eds), *Fuzzy Antomata and Decision Processes*, New York, North-Holland, 1977.

[R5.26] Gupta, M. M. and E. Sanchez (eds.), *Fuzzy Information and Decision Processes*, New York, North-Holland, 1982.

[R5.27] Higashi, M. and G. J. Klir, "On measure of fuzziness and fuzzy complements," *Intern. J. of General Systems*, Vol. 8 #3, 1982, pp. 169–180.

[R5.28] Higashi, M. and G. J. Klir, "Measures of uncertainty and information based on possibility distributions," *International Journal of General Systems*, Vol. 9 #1, 1983, pp. 43–58.

[R5.29] Higashi, M. and G. J. Klir, "On the notion of distance representing information closeness: Possibility and probability distributions," *Intern. J. of General Systems*, Vol. 9 #2, 1983, pp. 103–115.

[R5.30] Higashi, M. and G. J. Klir, "Resolution of finite fuzzy relation equations," *Fuzzy Sets and Systems*, Vol. 13, #1,1984, pp. 65–82.

[R5.31] Higashi, M. and G. J. Klir, "Identification of fuzzy relation systems," *IEEE Trans. on Systems, Man, and Cybernetics*, Vol. 14 #2, 1984, pp. 349–355.

[R5.32] Jin-wen, Z., "A Unified Treatment of Fuzzy Set Theory and Boolean Valued Set theory: Fuzzy Set Structures and Normal Fuzzy Set Structures," *Jour. Math. Anal. and Applications*, Vol. 76, #1, 1980, pp. 197–301.

[R5.33] Kandel, A. and W. J. Byatt, "Fuzzy Processes," *Fuzzy Sets and Systems*, Vol. 4, #2, 1980, pp. 117–152.

[R5.34] Kaufmann, A. and M. M. Gupta, *Introduction to fuzzy arithmetic: Theory and applications*, New York, Van Nostrand Reinhold, 1991.

[R5.35] Kaufmann, A., *Introduction to the Theory of Fuzzy Subsets*, Vol. 1, New York, Academic Press, 1975.

[R5.36] Klement, E. P. and W. Schwyhla, "Correspondence Between Fuzzy Measures and Classical Measures," *Fuzzy Sets and Systems*, Vol. 7, #1, 1982. pp. 57–70.

[R5.37] Klir, George and Bo Yuan, *Fuzzy Sets and Fuzzy Logic*, Upper Saddle River, NJ Prentice Hall, 1995.

[R5.38] Kruse, R. et al., *Foundations of Fuzzy Systems*, New York, Wiley, 1994.

[R5.37] Lasker, G. E. (ed.), *Applied Systems and Cybernetics, Vol. VI: Fuzzy Sets and Systems*, Pergamon Press, New York, 1981.

[R5.38] Lientz, B. P., "On Time Dependent Fuzzy Sets", *Inform, Science*, Vol. 4, 1972, pp. 367–376.

[R5.39] Lowen, R., "Fuzzy Uniform Spaces," *Jour. Math. Anal. Appl.*, Vol. 82, #21981, pp. 367–376.

[R5.40] Michalek, J., "Fuzzy Topologies," *Kybernetika*, Vol. 11, 1975, pp. 345–354.

[R5.41] Negoita, C. V. et al., *Applications of Fuzzy Sets to Systems Analysis*, Wiley and Sons, New York, 1975.

[R5.42] Negoita, C. V., "Representation Theorems for Fuzzy Concepts," *Kybernetes*, Vol. 4, 1975, pp. 169–174.

[R5.43] Negoita, C. V. et al., "On the State Equation of Fuzzy Systems," *Kybernetes*, Vol. 4, 1975, pp. 231–214.

[R5.44] Netto, A. B., "Fuzzy Classes," *Notices, American Mathematical Society*, Vol. 68T-H28, 1968, p. 945.

[R5.45] Pedrycz, W., "Fuzzy Relational Equations with Generalized Connectives and Their Applications," *Fuzzy Sets and Systems*, Vol. 10, #2, 1983, pp. 185–201.

[R5.46] Raha, S. et al., "Analogy Between Approximate Reasoning and the Method of Interpolation," *Fuzzy Sets and Systems*, Vol. 51, #3, 1992, pp. 259–266.

[R5.47] Ralescu, D., "Toward a General Theory of Fuzzy Variables," *Jour. of Math. Analysis and Applications*, Vol. 86, #1, 1982, pp. 176–193.

[R5.48] Rodabaugh, S. E., "Fuzzy Arithmetic and Fuzzy Topology," in G.E. Lasker, (ed.), *Applied Systems and Cybernetics, Vol. VI: Fuzzy Sets and Systems*, Pergamon Press, New York, 1981, pp. 2803–2807.

[R5.49] Rosenfeld, A., "Fuzzy Groups," *Jour. Math. Anal. Appln.*, Vol. 35, 1971, pp. 512–517.

[R5.50] Ruspini, E. H., "Recent Developments In Mathematical Classification Using Fuzzy Sets," in G.E. Lasker, (ed.), *Applied Systems and Cybernetics, Vol. VI: Fuzzy Sets and Systems*, Pergamon Press, New York, 1981. pp. 2785–2790.

[R5.51] Santos, E. S., "Fuzzy Algorithms," *Inform. and Control*, Vol. 17, 1970, pp. 326–339.

[R5.52] Stein, N. E. and K. Talaki, "Convex Fuzzy Random Variables," *Fuzzy Sets and Systems*, Vol. 6, #3, 1981, pp. 271–284.

[R5.53] Triantaphyllon, E. et al., "The Problem of Determining Membership Values in Fuzzy Sets in Real World Situations," in D.E. Brown et al. (eds), *Operations Research and Artificial Intelligence: The Integration of Problem-solving Strategies*, Boston, Mass., Kluwer, 1990, pp. 197–214.

[R5.54] Tsichritzis, D., "Participation Measures," *Jour. Math. Anal. and Appln.*, Vol. 36, 1971, pp. 60–72.

[R5.55] Turksens, I. B., "Four Methods of Approximate Reasoning with Interval-Valued Fuzzy Sets," *Intern. Journ. of Approximate Reasoning*, Vol. 3, #2, 1989, pp. 121–142.

[R5.56] Turksen, I. B., "Measurement of Membership Functions and Their Acquisition," *Fuzzy Sets and Systems*, Vol. 40, #1, 1991, pp. 5–38.

[R5.57] Wang, P. P. (ed.), *Advances in Fuzzy Sets, Possibility Theory, and Applications*, New York, Plenum Press, 1983.

[R5.58] Wang, Zhenyuan, and George Klir, *Fuzzy Measure Theory*, New York, Plenum Press, 1992.

[R5.59] Wang, P. Z. et al. (eds.), *Between Mind and Computer: Fuzzy Science and Engineering*, Singapore, World Scientific Press, 1993.

[R5.60] Wang, S., "Generating Fuzzy Membership Functions: A Monotonic Neural Network Model," *Fuzzy Sets and Systems*, Vol. 61, #1, 1994, pp. 71–82.

[R5.61] Wong, C. K., "Fuzzy Points and Local Properties of Fuzzy Topology," *Jour. Math. Anal. and Appln.*, Vol. 46, 19874, pp. 316–328.
[R5.62] Wong, C. K., "Categories of Fuzzy Sets and Fuzzy Topological Spaces," *Jour. Math. Anal. and Appln.*, Vol. 53, 1976, pp. 704–714.
[R5.62] Yager, R. R. et al., (Eds.), *Fuzzy Sets, Neural Networks, and Soft Computing*, New York, Nostrand Reinhold, 1994.
[R5.62] Zadeh, L. A., "A Computational Theory of Decompositions," *Intern. Jour. of Intelligent Systems*, Vol. 2, #1, 1987, pp. 39–63.
[R5.63] Zimmerman, H. J., *Fuzzy Set Theory and Its Applications*, Boston, Mass, Kluwer, 1985.

R6. Fuzzy Optimization, Information and Decision-Choice Sciences

[R6.1] Bose, R. K. and Sahani D, "Fuzzy Mappings and Fixed Point Theorems," *Fuzzy Sets and Systems*, Vol. 21, 1987, pp. 53–58.
[R6.2] Butnariu D. "Fixed Points for Fuzzy Mappings," *Fuzzy Sets and Systems*, Vol. 7, 1982, pp. 191–207.
[R6.3] Dompere, Kofi K., "Fuzziness, Rationality, Optimality and Equilibrium in Decision and Economic Theories" in Weldon A. Lodwick and Janusz Kacprzyk (Eds.), *Fuzzy Optimization: Recent Advances and Applications* (Series: Studies in Fuzziness and Soft Computing, Vol. 254), Berlin, Heidelberg, Springer, 2010.
[R6.4] Eaves, B. C., "Computing Kakutani Fixed Points," *Journal of Applied Mathematics*, Vol. 21, 1971, pp. 236–244.
[R6.5] Heilpern, S. "Fuzzy Mappings and Fixed Point Theorem," *Journal of Mathematical Analysis and Applications*, Vol. 83, 1981, pp. 566–569.
[R6.6] Kacprzyk, J. et al., (eds.), *Optimization Models Using Fuzzy Sets and Possibility Theory*, Boston, Mass., D. Reidel, 1987.
[R6.7] Kaleva, O. "A Note on Fixed Points for Fuzzy Mappings", *Fuzzy Sets and Systems*, Vol. 15, 1985, pp. 99–100.
[R6.8] Lodwick, Weldon A and Janusz Kacprzyk (eds.), *Fuzzy Optimization: Recent Advances and Applications*, (Studies in Fuzziness and Soft Computing, Vol. 254), Berlin Heidelberg, Springer, 2010.
[R6.9] Negoita, C. V., "The Current Interest in Fuzzy Optimization," *Fuzzy Sets and Systems*, Vol. 6, #3, 1981, pp. 261–270.
[R6.10] Negoita, C. V., et al., "On Fuzzy Environment in Optimization Problems," in J. Rose et al., (eds.), *Modern Trends in Cybernetics and Systems,* Springer, Berlin, 1977, pp. 13–24.
[R6.11] Zimmerman, H.-J., "Description and Optimization of Fuzzy Systems," *Intern. Jour. Gen. Syst*. Vol. 2, #4, 1975, pp. 209–215.

R7. Ideology, Disinformation, Misinformation and Propaganda

[R7.1] Abercrombie, Nicholas et al., *The Dominant Ideology Thesis*, London, Allen and Unwin, 1980.
[R7.2] Abercrombie, Nicholas, *Class, Structure, and Knowledge: Problems in the Sociology of Knowledge*, New York, New York University Press, 1980.
[R7.3] Aron, Raymond, *The Opium of the Intellectuals*, Lanham, MD, University Press of America, 1985.
[R7.4] Aronowitz, Stanley, *Science as Power: Discourse and Ideology in Modern Society,* Minneapolis, University of Minnesota Press, 1988.

[R7.5] Barinaga, M. and E. Marshall, *Confusion on the Cutting Edge*, Science, Vol. 257, July 1992, pp. 616–625.
[R7.6] Barnett, Ronald, *Beyond All Reason: Living with Ideology in the University,* Philadelphia, PA., Society for Research into Higher Education and Open University Press, 2003.
[R7.7] Barth, Hans, *Truth and Ideology*, Berkeley, University of California Press, 1976.
[R7.8] Basin, Alberto, and Thierry Verdie, "The Economics of Cultural Transmission and the Dynamics of Preferences," *Journal of Economic Theory*, Vol. 97, 2001, pp. 298–319.
[R7.9] Bikhchandani, Sushil et al., "A Theory of Fads, Fashion, Custom, and Cultural Change," *Journal of political Economy*, Vol. 100 1992, pp. 992–1026.
[R7.10] Boyd Robert and Peter J Richerson, *Culture and Evolutionary Process*, Chicago, University of Chicago Press, 1985.
[R7.11] Buczkowski, Piotr and Andrzej Klawiter, *Theories of Ideology and Ideology of Theories*, Amsterdam, Rodopi, 1986.
[R7.12] Chomsky, Norm, *Manufacturing Consent*, New York, Pantheo Press, 1988.
[R7.13] Chomsky, N., *Problem of Knowledge and Freedom*, Glasgow, Collins, 1972.
[R7.14] Cole, Jonathan, R., "Patterns of Intellectual influence in Scientific Research," *Sociology of Education*, Vol. 43, 1968, pp. 377–403.
[R7.15] Cole Jonathan, R. and Stephen Cole, *Social Stratification in Science*, Chicago, University of Chicago Press, 1973.
[R7.16] Debackere, Koenraad and Michael A. Rappa, "Institutioal Variations in Problem Choice and Persistence among Scientists in an Emerging Fields," *Research Policy,* Vol. 23, 1994, pp. 425–441.
[R7.17] Fraser, Colin and George Gaskell (eds.), *The Social Psychological Study of Widespread Beliefs*, Oxford, Clarendon Press, 1990.
[R7.18] Gieryn, Thomas, F. "Problem Retention and Problem Change in Science," *Sociological Inquiry*, Vol. 48, 1978, pp. 96–115.
[R7.19] Harrington, Joseph E. Jr, "The Rigidity of social Systems," *Journal of Political Economy*, Vol. 107, pp. 40–64.
[R7.20] Hinich, Melvin and Michael Munger, *Ideology and the Theory of Political Choice*, Ann Arbor University of Michigan Press, 1994.
[R7.21] Hull, D. L., *Science as a Process: An Evolutionary Account of the Social and Conceptual Development of Science,* Chicago, University of Chicago Press, 1988.
[R7.22] Marx, Karl and Friedrich Engels, *The German Ideology*, New York, International Pub., 1970.
[R7.23] Mészáros István, *Philosophy, Ideology and Social Science*: Essay in Negation and Affirmation, Brighton, Sussex, Wheatsheaf, 1986.
[R7.24] Mészáros István, *The Power of Ideology*, New York, New York University Press, 1989.
[R7.25] Newcomb, Theodore M. et al., *Persistence and Change*, New York, Wiley, 1967.
[R7.26] Pickering, Andrew, *Science as Practice and Culture,* Chicago, University of Chicago Press, 1992.
[R7.27] Therborn, Göran, *The Ideology of Power and the Power of Ideology*, London, NLB Publications, 1980.
[R7.28] Thompson, Kenneth, *Beliefs and Ideology*, New York, Tavistock Publication, 1986.
[R7.29] Ziman, John, "The Problem of 'Problem Choice'," *Minerva*, Vol. 25, 1987, pp. 92–105.
[R7.30] Ziman, John, *Public Knowledge: An Essay Concerning the Social Dimension of Science*, Cambridge, Cambridge University Press, 1968.
[R7.31] Zuckerman, Hrriet, "Theory Choice and Problem Choice in Science," *Sociological Inquiry*, Vol. 48, 1978, pp. 65–95.

R8. Information, Thought and Knowledge

[R8.1] Aczel, J. and Z. Daroczy, *On Measures of Information and their Characterizations*, New York, Academic Press, 1975.
[R8.2] Afanasyev, Social Information and Regulation of Social Development, Moscow, Progress, 1878.
[R8.3] Anderson, J. R., *The Architecture of Cognition*, Cambridge, Mass., Harvard University Press, 1983.
[R8.4] Angelov, Stefan and Dimitr Georgiev, "The Problem of Human Being in Contemporary Scientic Knowledge," *Soviet Studies in Philosophy*, Summer, 1974, pp. 49–66.
[R8.5] Ash, Robert, *Information Theory*, New York, Wiley, 1965.
[R8.6] Bergin, J., "Common Knowledge with Monotone Statistics," *Econometrica*, Vol. 69, 2001, pp. 1315–1332.
[R8.7] Bestougeff, Hélène and Gerard Ligozat, *Logical Tools for Temporal Knowledge Representation*, New York, Ellis Horwood, 1992.
[R8.8] Brillouin, L., *Science and information Theory*, New York, Academic Press, 1962.
[R8.9] Bruner, J. S., et al., *A Study of Thinking*, New York, Wiley, 1956.
[R8.10] Brunner, K. and A. H. Meltzer (eds.), *Three Aspects of Policy and Policy Making: Knowledge, Data and Institutions*, Carnegie-Rochester Conference Series, Vol. 10, Amsterdam, North-Holland, 1979.
[R8.11] Burks, A. W., *Chance, Cause, Reason: An Inquiry into the Nature of Scientific Evidence*, Chicago, University of Chicago Press, 1977.
[R8.12] Calvert, Randall, *Models of Imperfect Information in Politics*, New York, Hardwood Academic Publishers, 1986.
[R8.13] Cornforth, Maurice, *The Theory of Knowledge*, New York, International Pub., 1972.
[R8.14] Cornforth, Maurice, *The Open Philosophy and the Open Society*, New York, International Pub., 1970.
[R8.15] Coombs, C. H., *A Theory of Data*, New York, Wiley, 1964.
[R8.16] Dretske, Fred. I., *Knowledge and the Flow of Information*, Cambridge, Mass., MIT Press, 1981.
[R8.17] Dreyfus, Hubert L., "A Framework for Misrepresenting Knowledge," in Martin Ringle (ed.) *Philosophical Perspectives in Artificial Intelligence*, Atlantic Highlands, N.J., Humanities Press, 1979.
[R8.18] Fagin R. et al., *Reasoning About Knowledge*, Cambridge, Mass, MIT Press, 1995.
[R8.19] Geanakoplos, J., "Common Knowledge," *Journal of Economic Perspectives*," Vol. 6, 1992, pp. 53–82.
[R8.20] George, F. H., *Models of Thinking*, London, Allen and Unwin, 1970.
[R8.21] George, F. H., "Epistemology and the problem of perception," *Mind*, Vol. 66, 1957, pp. 491–506.
[R8.22] Harwood, E. C., *Reconstruction of Economics*, Great Barrington, Mass, American Institute for Economic Research, 1955.
[R8.23] Hintikka, J., *Knowledge and Belief*, Ithaca, N.Y., Cornell University Press, 1962.
[R8.24] Hirshleifer, Jack., "The Private and Social Value of Information and Reward to inventive activity," *American Economic Review*, Vol. 61, 1971, pp. 561–574.
[R8.25] Kapitsa, P. L., "The Influence of Scientific Ideas on Society," *Soviet Studies in Philosophy*, Fall, 1979, pp. 52–71.
[R8.26] Kedrov, B. M., "The Road to Truth," *Soviet Studies in Philosophy*, Vol. 4, 1965, pp. 3–53.
[R8.27] Klatzky, R. L., *Human Memory: Structure and Processes*, San Francisco, Ca., W. H. Freeman Pub., 1975.
[R8.28] Kreps, David and Robert Wilson, "Reputation and Imperfect Information," *Journal of Economic Theory*, Vol. 27. 1982, pp. 253–279.
[R8.29] Kubát, Libor and J. Zeman (eds.), *Entropy and Information*, Amsterdam, Elsevier, 1975.

[R8.30] Kurcz, G. and W. Shugar et al (eds.), *Knowledge and Language*, Amsterdam, North-Holland, 1986.

[R8.31] Lakemeyer, Gerhard,and Bernhard Nobel (eds.), *Foundations of Knowledge Representation and Reasoning*, Berlin, Springer, 1994.

[R8.32] Lektorskii, V. A., "Principles involved in the Reproduction of Objective in Knowledge,", *Soviet Studies in Philosophy*, Vol. 4, #4, 1967, pp. 11–21.

[R8.33] Levi, I., *The Enterprise of Knowledge*, Cambridge, Mass. MIT Press, 1980.

[R8.34] Levi, Isaac, "Ignorance, Probability and Rational Choice", *Synthese*, Vol. 53, 1982, pp. 387–417.

[R8.35] Levi, Isaac, "Four Types of Ignorance," *Social Science*, Vol. 44, pp. 745–756.

[R8.36] Marschak, Jacob, *Economic Information, Decision and Prediction: Selected Essays*, Vol. II, Part II, Boston, Mass. Dordrecht-Holland, 1974.

[R8.37] Menges, G. (ed.), *Information, Inference and Decision*, D. Reidel Pub., Dordrecht, Holland, 1974.

[R8.38] Michael Masuch and László Pólos (eds.), *Knowledge Representation and Reasoning Under Uncertainty*, New York, Springer, 1994.

[R8.39] Moses, Y. (ed.), *Proceedings of the Fourth Conference of Theoretical Aspects of Reasoning about Knowledge*, San Mateo, Morgan Kaufmann, 1992.

[R8.40] Nielsen, L. T. et al., "Common Knowledge of Aggregation Expectations," *Econometrica*, Vol. 58, 1990, pp. 1235–1239.

[R8.41] Newell, A., *Unified Theories of Cognition*, Cambridge, Mass. Harvard University Press, 1990.

[R8.42] Newell, A., *Human Problem Solving*, Englewood Cliff, N.J., Prentice-Hall, 1972.

[R8.43] Ogden, G. K. and I. A., *The Meaning of Meaning*, New York, Harcourt-Brace Jovanovich, 1923.

[R8.44] Planck, Max, Scientific Autobiography and Other Papers, Westport, Conn., Greenwood, 1968.

[R8.45] Pollock, J., *Knowledge and Justification*, Princeton, Princeton University Press, 1974.

[R8.46] Polanyi, M., *Personal Knowledge*, London, Routledge and Kegan Paul, 1958.

[R8.47] Popper, K. R., *Objective Knowledge*, London, Macmillan, 1949.

[R8.48] Popper, K. R., *Open Society and it Enemies, Vols. 1 and 2* Princeton, Princeton Univ. Press, 2013.

[R8.49] Popper, K. R., *The Poverty of Historicism* New York, Taylor and Francis, 2002.

[R8.50] Price, H. H., *Thinking and Experience*, London, Hutchinson, 1953.

[R8.51] Putman, H., *Reason, Truth and History*, Cambridge, Cambridge University Press, 1981.

[R8.52] Putman. H., *Realism and Reason*, Cambridge, Cambridge University Press, 1983.

[R8.53] Putman, H., *The Many Faces of Realism*, La Salle, Open Court Publishing Co., 1987.

[R8.54] Russell, B., *Human Knowledge, its Scope and Limits*, London, Allen and Unwin, 1948.

[R8.55] Russell, B., *Our Knowledge of the External World,* New York, Norton, 1929.

[R8.56] Samet, D., "Ignoring Ignorance and Agreeing to Disagree," *Journal of Economic Theory*, Vol. 52, 1990, pp. 190–207.

[R8.57] Schroder, Harold, M. and Peter Suedfeld (eds.), *Personality Theory and Information Processing*, New York, Ronald Pub. 1971.

[R8.58] Searle J., *Minds, Brains and Science*, Cambridge, Mass., Harvard University Press, 1985.

[R8.59] Shin, H., "Logical Structure of Common Knowledge," *Journal of Economic Theory*, Vol. 60, 1993, pp. 1–13.

[R8.60] Simon, H. A., *Models of Thought*, New Haven, Conn., Yale University Press, 1979.

[R8.61] Smithson, M., *Ignorance and Uncertainty, Emerging Paradigms*, New York, Springer, 1989.

[R8.62] Sowa, John F., *Knowledge Representation: Logical, Philosophical, and Computational Foundations*, Pacific Grove, Brooks Pub., 2000.

[R8.63] Stigler, G. J., The Economics of Information," *Journal of Political Economy*, Vol.69, 1961, pp. 213–225.

[R8.64] Tiukhtin, V. S., "How Reality Can be Reflected in Cognition: Reflection as a Property of All Matter," *Soviet Studies in Philosophy*, Vol. 3 #1, 1964, pp. 3–12.
[R8.65] Tsypkin, Ya Z., *Foundations of the Theory of Learning Systems*, New York, Academic Press, 1973.
[R8.66] Ursul, A. D., "The Problem of the Objectivity of Information," in Libor Kubát, and J. Zeman (eds.), *Entropy and Information*, Amsterdam, Elsevier, 1975. pp. 187–230.
[R8.67] Vardi, M. (ed.), *Proceedings of Second Conference on Theoretical Aspects of Reasoning about Knowledge*, Asiloman, Ca., Los Altos, Ca, Morgan Kaufman, 1988.
[R8.68] Vazquez, Mararita, et al., "Knowledge and Reality: Some Conceptual Issues in System Dynamics Modeling," *Systems Dynamics Review,* Vol. 12, 1996, pp. 21–37.
[R8.69] Zadeh, L. A., "A Theory of Commonsense Knowledge," in Skala, Heinz J. et al., (eds.), *Aspects of Vagueness*, Dordrecht, D. Reidel Co. 1984, pp. 257–295.
[R8.70] Zadeh, L. A., "The Concept of Linguistic Variable and its Application to Approximate reasoning," *Information Science*, Vol. 8, 1975, pp. 199–249 (Also in Vol. 9, pp. 40–80).

R9. Languages and Information

[R9.1] Agha, Agha, *Language and Social Relations,* Cambridge, Cambridge University Press, 2006.
[R9.2] Aitchison, Jean (ed.), *Language Change: Progress or Decay?* Cambridge, New York, Melbourne: Cambridge University Press, 2001.
[R9.3] Anderson, Stephen, *Languages: A Very Short Introduction.* Oxford: Oxford University Press (2012).
[R9.4] Aronoff, Mark and Fudeman, Kirsten, *What is Morphology.* New York, Wiley, 2011.
[R9.5] Bauer, Laurie (ed.), *Introducing linguistic morphology* Washington, D.C.: Georgetown University Press, 2003.
[R9.6] Barber Alex and Robert J Stainton (eds.). *Concise Encyclopedia of Philosophy of Language and Linguistics.* New York, Elsevier, 2010.
[R9.7] Brown, Keith; Ogilvie, Sarah, (eds.), *Concise Encyclopedia of Languages of the World,* New York, Elsevier Science, 2008.
[R9.8] Campbell, Lyle (ed.), *Historical Linguistics: an Introduction Cambridge*, MASS, MIT Press, 2004.
[R9.9] Chomsky, Noam, *Syntactic Structures.* The Hague: Mouton, 1957.
[R9.10] Chomsky, Noam, *The Architecture of Language.* Oxford, Oxford University Press, 2000.
[R9.11] Clarke, David S. *Sources of semiotic: readings with commentary from antiquity to the present* Carbondale: Southern Illinois University Press, 1990.
[R9.12] Collinge, N. E. (ed.), *An Encyclopedia of Language.* London: New York: Routledge (1989).
[R9.13] Comrie, Bernard (ed.), *Language universals and linguistic typology: Syntax and morphology,* Oxford, Blackwell, 1989.
[R9.14] Comrie, Bernard (ed.), *The World's Major Languages.* New York: Routledge, 2009.
[R9.15] Coulmas, Florian, *Writing Systems: An Introduction to Their Linguistic Analysis.* Cambridge, Cambridge University Press, 2002.
[R9.16] Croft, William, Cruse, D. Alan, *Cognitive Linguistics.* Cambridge, Cambridge University Press, 2004.
[R9.17] Croft, William, "Typology". In Mark Aronoff; Janie Rees-Miller (ed.), *The Handbook of Linguistics*, Oxford, Blackwell, pp. 81–105, 2001.
[R9.18] Crystal, David (ed.) *The Cambridge Encyclopedia of Language.* Cambridge: Cambridge University Press, 1997.
[R9.20] Deacon, Terrence, *The Symbolic Species: The Co-evolution of Language and the Brain,* New York: W.W. Norton & Company, 1997.

[R9.21] Devitt, Michael and Sterelny, Kim, *Language and Reality: An Introduction to the Philosophy of Language.* Boston: MIT Press, 1999.

[R9.22] Duranti, Alessandro "Language as Culture in U.S. Anthropology: Three Paradigms" *Current Anthropology*, Vol. **44** (3), pp. 323–348, 2003.

[R9.23] Evans, Nicholas and Levinson, Stephen C, "The myth of language universals: Language diversity and its importance for cognitive science," Vol. **32** (5). *Behavioral and Brain Sciences,* pp. 429–492, 2009.

[R9.24] Fitch, W. Tecumseh, *The Evolution of Language,* Cambridge, Cambridge University Press, 2010.

[R9.25] Foley, William A., *Anthropological Linguistics: An Introduction,* Oxford, Blackwell, 1997.

[R9.26] Ginsburg, Seymour, *Algebraic and Automata-Theoretic Properties of Formal Languages,* New York, North-Holland, 1973.

[R9.27] Goldsmith, John A, *The Handbook of Phonological Theory: Blackwell Handbooks in Linguistics.* Oxford, Blackwell Publishers, 1995.

[R9.28] Greenberg, Joseph, *Language Universals: With Special Reference to Feature Hierarchies.* The Hague, Mouton & Co., 1966.

[R9.29] Hauser, Marc D.; Chomsky, Noam; Fitch, W. Tecumseh, "The Faculty of Language: What Is It, Who Has It, and How Did It Evolve?" *Science,* 22 **298** (5598), pp. 1569–1579, 2002.

[R9.30] Hörz,Herbert, "Information, Sign, Image," in Libor Kubát and Jiří Zeman (eds.) *Entropy and Information in Science and Philosophy,* New York, Elsevier, 1975.

[R9.31] International Phonetic Association, *Handbook of the International Phonetic Association: A guide to the use of the International Phonetic Alphabet.* Cambridge, Cambridge University Press (1999).

[R9.32] Katzner, Kenneth, *The Languages of the World,* New York: Routledge, 1999.

[R9.33] Labov, William, *Principles of Linguistic Change vol. I Internal Factors,* Oxford, Blackwell, 1994.

[R9.34] Labov, William, *Principles of Linguistic Change vol. II Social Factors*, Oxford, Blackwell, 2001.

[R9.35] Levinson, Stephen C. *Pragmatics.* Cambridge: Cambridge University Press, 1983.

[R9.36] Lewis, M. Paul (ed.), *Ethnologue: Languages of the World,* Dallas, Tex.: SIL International, 2009.

[R9.37] Lyons, John, *Language and Linguistics,* Cambridge, Cambridge University Press, 1981.

[R9.38] MacMahon, *April M.S., Understanding Language Change,* Cambridge, Cambridge University Press, 1994.

[R9.39] Matras, Yaron; Bakker, Peter, (eds). *The Mixed Language Debate: Theoretical and Empirical Advances.* Berlin: Walter de Gruyter, 2003.

[R9.40] Moseley, Christopher (ed), *Atlas of the World's Languages in Danger,* Paris: UNESCO Publishing (2010).

[R9.41] Nerlich, B. "History of pragmatics". In L. Cummings (ed.), *The Pragmatics Encyclopedia.* New York: Routledge, pp. 192–193, 2010.

[R9.42] Newmeyer, Frederick J., *The History of Linguistics*. Linguistic Society of America, 2005.

[R9.43] Newmeyer, Frederick J., *Language Form and Language Function* (PDF). Cambridge, MA: MIT Press, 1998.

[R9.44] Nichols, Johanna, *Linguistic diversity in space and time.* Chicago, University of Chicago Press, 1992.

[R9.45] Nichols, Johanna. "*Functional Theories of Grammar*". *Annual Review of Anthropology,* Vol. **13**: 19849, pp. 7–117.

[R9.46] Sandler, Wendy; Lillo-Martin, Diane, "Natural Sign Languages", In Mark Aronoff; Janie Rees-Miller (eds.). *The Handbook of Linguistics.* Oxford, Blackwell, pp. 533–563, 2001.

[R9.47] Swadesh, Morris, "*The phonemic principle*", *Language,* Vol. **10** (2): (1934), pp. 117–129.

[R9.48] Tomasello, Michael, "The Cultural Roots of Language". In B. Velichkovsky and D. Rumbaugh (eds.), *Communicating Meaning: The Evolution and Development of Language.* New York, Psychology Press, pp. 275–308, 1996.
[R9.49] Tomasello, Michael, *Origin of Human Communication.* Cambridge Mass., MIT Press, 2008.
[R9.50] Thomason, Sarah G, *Language Contact– An Introduction,* Edinburgh, Edinburgh University Press, 2001.
[R9.51] Ulbaek, Ib, "The Origin of Language and Cognition", In J. R. Hurford and C. Knight (eds.). *Approaches to the evolution of language.* Cambridge, Cambridge University Press. 1998, pp. 30–43.
[R9.52] Van Valin, Jr, Robert D., "Functional Linguistics," In Mark Aronoff; Janie Rees-Miller (eds.), *The Handbook of Linguistics.* Oxford, Blackwell (2001), pp. 319–337.

R10. Language, Knowledge-Production Process and Epistemics

[R10.1] Aho, A. V. "Indexed Grammar - An Extension of Context-Free Grammars" *Journal of the Association for Computing Machinery*, Vol. 15, 1968, pp. 647–671.
[R10.2] Black, Max (ed.), *The Importance of Language*, Englewood Cliffs, NJ, Prentice-Hall, 1962.
[R10.3] Carnap, Rudolff, Meaning and Necessity: A Study in Semantics and Modal Logic, Chicago, University of Chicago Press, 1956.
[R10.4] Chomsky, Norm, "Linguistics and Philosophy" in S. Hook (ed.) *Language and Philosophy*, New York, New York University Press, 1968, pp. 51–94.
[R10.5] Chomsky, Norm, *Language and Mind*, New York, Harcourt Brace Jovanovich, 1972.
[R10.6] Cooper, William S., *Foundations of Logico-Linguistics: A Unified Theory of Information, Language and Logic*, Dordrecht, D. Reidel, 1978.
[R10.7] Cresswell, M. J.., *Logics and Languages*, London, Methuen Pub., 1973.
[R10.8] Dilman, Ilham, *Studies in Language and Reason*, Totowa, N.J., Barnes and Nobles, Books, 1981.
[R10.9] Fodor, Jerry A., *The Language and Thought*, New York, Thom as Y. Crowell Co, 1975.
[R10.10] Givon, Talmy, *On Understanding Grammar*, New York, Academic Press, 1979.
[R10.11] Gorsky, D. R., *Definition*, Moscow, Progress Publishers, 1974.
[R10.12] Hintikka, Jaakko, The Game of Language, Dordrecht, D. Reidel Pub., 1983.
[R10.13] Johnson-Lair, Philip N. *Mental Models: Toward Cognitive Science of Language, Inference and Consciousness*, Cambridge, Mass, Harvard University Press, 1983.
[R10.14] Kandel, A., "Codes Over Languages," *IEEE Transactions on Systems Man and Cybernetics*, Vol. 4, 1975, pp. 135–138.
[R10.15] Keenan, Edward L. and Leonard M. Faltz, *Boolean Semantics for Natural Languages*, Dordrecht, D. Reidel Pub., 1985.
[R10.16] Lakoff, G. Linguistics and Natural Logic, *Synthese*, Vol. 22, 1970, pp. 151–271.
[R10.17] Lee, E. T., et al., "Notes On Fuzzy Languages," *Information Science*, Vol. 1, 1969, pp. 421–434.
[R10.18] Mackey, A. and D. Merrill (eds.) *Issues in the Philosophy of Language*, New Haven, CT, Yale University Press, 1976.
[R10.19] Nagel, T., "Linguistics and Epistemology" in S. Hook (ed.) *Language and Philosophy*, New York, New York University Press, 1969, pp. 180–184.
[R10.20] Pike, Kenneth, *Language in Relation to a Unified Theory of Structure of Human Behavior*, The Hague, Mouton Pub., 1969.
[R10.21] Quine, W. V. O. *Word and object*, Cambridge, Mass, MIT Press, 1960.
[R10.22] Russell, Bernard, *An Inquiry into Meaning and Truth*, Penguin Books, 1970.
[R10.23] Tarski, Alfred, *Logic, Semantics and Mathematics*, Oxford, Clarendon Press, 1956.
[R10.24] Whorf, B. L. (ed.), *Language, Thought and Reality*, New York, Humanities Press, 1956.

R11. Possible-Actual Worlds and Information Analytics

[R11.1] Adams, Robert M., "Theories of Actuality," *Noûs*, Vol. 8, 1974, pp. 211–231.

[R11.2] Allen, Sture (ed.) *Possible Worlds in Humanities, Arts and Sciences*, Proceedings of Nobel Symposium, Vol. 65, New York, Walter de Gruyter Pub., 1989.

[R11.3] Armstrong, D. M., *A Combinatorial Theory of Possibility.* Cambridge University Press, 1989.

[R11.4] Armstrong, D. M., *A World of States of Affairs*, Cambridge, Cambridge University Press, 1997.

[R11.5] Bell, J. S., "Six Possible Worlds of Quantum Mechanics" in Allen, Sture (Ed.) *Possible Worlds in Humanities, Arts and Sciences*, Proceedings of Nobel Symposium, Vol. 65, New York, Walter de Gruyter Pub., 1989, pp. 359–373.

[R11.6] Bigelow, John. "Possible Worlds Foundations for Probability", *Journal of Philosophical Logic,* 5 (1976), pp. 299–320.

[R11.7] Bradley, Reymond and Norman Swartz, *Possible World: An Introduction to Logic and its Philosophy,* Oxford, Bail Blackwell, 1997.

[R11.8] Castañeda, H.-N. "Thinking and the Structure of the World", *Philosophia*, 4 (1974), pp. 3–40.

[R11.9] Chihara, Charles S. *The Worlds of Possibility: Modal Realism and the Semantics of Modal Logic*, Clarendon, 1998.

[R11.10] Chisholm, Roderick. "Identity through Possible Worlds: Some Questions", *Noûs,* 1 (1967), pp. 1–8; reprinted in Loux, *The Possible and the Actual.*

[R11.11] Divers, John, *Possible Worlds*, London: Routledge, 2002.

[R11.12] Forrest, Peter. "Occam's Razor and Possible Worlds", *Monist,* 65 (1982), pp. 456–464.

[R11.13] Forrest, Peter, and Armstrong, D. M. "An Argument Against David Lewis' Theory of Possible Worlds", *Australasian Journal of Philosophy,* 62 (1984), pp. 164–168.

[R11.14] Grim, Patrick, "There is No Set of All Truths", *Analysis*, Vol. 46, 1986, pp. 186–191.

[R11.15] Heller, Mark. "Five Layers of Interpretation for Possible Worlds", *Philosophical Studies*, 90 (1998), pp. 205–214.

[R11.16] Herrick, Paul, *The Many Worlds of Logic,* Oxford: Oxford University Press, 1999.

[R11.17] Krips, H. "Irreducible Probabilities and Indeterminism", *Journal of Philosophical Logic*, Vol. 18, 1989, pp. 155–172.

[R11.18] Kuhn, Thomas S., "Possible Worlds in History of Science" in Allen, Sture (ed.) *Possible Worlds in Humanities, Arts and Sciences,* Proceedings of Nobel Symposium, Vol. 65, New York, Walter de Gruyter Pub., 1989, pp. 9–41.

[R11.19] Kuratowski, K. and Mostowski, A. *Set Theory: With an Introduction to Descriptive Set Theory*, New York: North-Holland, 1976.

[R11.20] Lewis, David, *On the Plurality of Worlds*, Oxford, Basil Blackwell, 1986.

[R11.21] Loux, Michael J. (ed.) *The Possible and the Actual*: *Readings in the Metaphysics of Modality,* Ithaca & London: Cornell University Press, 1979.

[R11.22] Parsons, Terence, *Nonexistent Objects,* New Haven, Yale University Press, 1980.

[R11.23] Perry, John, "From Worlds to Situations", *Journal of Philosophical Logic*, Vol. 15, 1986, pp. 83–107.

[R11.24] Rescher, Nicholas and Brandom, Robert. *The Logic of Inconsistency: A Study in Non-Standard Possible-World Semantics And Ontology*, Rowman and Littlefield, 1979.

[R11.25] Skyrms, Brian. "Possible Worlds, Physics and Metaphysics", *Philosophical Studies,* Vol. 30, 1976, pp. 323–332.

[R11.26] Stalmaker, Robert C. "Possible World", *Noûs*, Vol. 10, 1976, pp. 65–75.

[R11.27] Quine, W. V. O. *Word and Object*, M.I.T. Press, 1960.

[R11.28] Quine, W. V. O "Ontological Relativity", *Journal of Philosophy*, 65 (1968), pp. 185–212.

R12. Philosophy of Information and Semantic Information

[R12.1] Aisbett, J., Gibbon, G.: "A practical measure of the information in a logical theory" Journal of Experimental and Theoretical Artificial Intelligence Vol. 11(2), 1999, pp. 201–218.
[R12.2] Badino, M.: "An Application of Information Theory to the Problem of the Scientific Experiment" *Synthese* Vol. 140, 2004, pp. 355–389.
[R12.3] Bar-Hillel, Y. (ed.), *Language and Information: Selected Essays on Their Theory and Application,* Reading, Addison-Wesley, (1964).
[R12.4] Bar-Hillel, Y., Carnap, R. "An Outline of a Theory of Semantic Information," (1953); in Bar-Hillel, Y. (ed.), *Language and Information: Selected Essays on Their Theory and Application,* Reading, Addison-Wesley, (1964), pp. 221–274.
[R12.5] Barwise, J., Seligman, J.: *Information Flow: The Logic of Distributed Systems*, Cambridge, University Press, Cambridge (1997).
[R12.6] Braman, S.: "Defining Information," *Telecommunications Policy*, Vol. 13, pp. 233–242 (1989).
[R12.7] Bremer, M. E.: "Do Logical Truths Carry Information?" *Minds and Machines* Vol. 13(4), 2003, pp. 567–575.
[R12.8] Bremer, M. and Cohnitz, D., *Information and Information Flow: an Introduction*, Ontos Verlag, Frankfurt, Lancaster, 2004.
[R12.9] Chaitin, G. J., *Algorithmic Information Theory*, Cambridge, Cambridge University Press, 1987.
[R12.10] Chalmers, D. J.: *The Conscious Mind: In Search of a Fundamental Theory*, New York, Oxford Univ. Press, (1996).
[R12.11] Cherry, C.: *On Human Communication: A Review, a Survey, and a Criticism*, Cambridge, MIT Press, 1978.
[R12.12] Colburn, T. R., *Philosophy and Computer Science*, Armonk, M.E. Sharpe, 2000.
[R12.13] Cover, T. M., Thomas, J. A.: *Elements of Information Theory*, New York, Wiley, 1991.
[R12.14] Dennett, D. C.: "Intentional Systems," *The Journal of Philosophy,* Vol. 68, 1971, pp. 87–106.
[R12.15] Deutsch, D., *The Fabric of Reality*, London Penguin, 1997.
[R12.16] Devlin, K. J., *Logic and Information*. Cambridge, Cambridge University Press, 1991.
[R12.17] Fetzer, J. H., "Information, Misinformation, and Disinformation" *Minds and Machines*, Vol. 14(2), 2004, pp. 223–229.
[R12.18] Floridi, L., *Philosophy and Computing: An Introduction*, London, Routledge, (1999).
[R12.19] Floridi, L., "What Is the Philosophy of Information?" *Metaphilosophy*, Vol. 33(1–2) (2002), pp. 123–145.
[R12.20] Floridi, L., "Two Approaches to the Philosophy of Information," *Minds and Machines,* Vol. 13(4), (2003), pp. 459–469.
[R12.21] Floridi, L.: "Open Problems in the Philosophy of Information," *Metaphilosophy*, Vol. 35 (4), 2004, pp. 554–582.
[R12.22] Floridi, L., "Outline of a Theory of Strongly Semantic Information," *Minds and Machines,* Vol. 14(2), 2004, pp. 197–222.
[R12.23] Floridi, L., "Is Information Meaningful Data?," *Philosophy and Phenomenological Research,* Vol. 70(2), 2005, pp. 351–370.
[R12.24] Fox, C. J.: *Information and Misinformation: An Investigation of the Notions of Information, Misinformation, Informing, and Misinforming,* Westport Greenwood Press, 1983.
[R12.25] Frieden, B. R., *Science from Fisher Information: A Unification*, Cambridge University Press, Cambridge, 2004.
[R12.26] Golan, A., "Information and Entropy Econometrics - Editor's View". Journal of Econometrics, Vol. 107(1–2), 2002, pp. 1–15.
[R12.27] Graham, G., *The Internet: A Philosophical Inquiry*, London, Routledge, 1999.
[R12.28] Grice, H. P. *Studies in the Way of Words,* Cambridge, Harvard University Press, 1989.

[R12.29] Hanson, P. P. (ed.), *Information, language, and cognition*, Vancouver, University of British Columbia Press, 1990.
[R12.30] Harms, W. F., "The Use of Information Theory in Epistemology," *Philosophy of Science*, Vol. 65(3), 472–501 (1998).
[R12.31] Heil, J., "Levels of Reality," *Ratio* Vol. 16(3), 2003, pp. 205–221.
[R12.32] Hintikka, J., Suppes, P. (eds.), *Information and Inference*, Reidel, Dordrecht, 1970.
[R12.33] Kemeny, J., "A Logical Measure Function," *Journal of Symbolic Logic*, Vol. 18, 1953, pp. 289–308.
[R12.34] Kolin K. K. "The Nature of Information and Philosophical Foundations of Informatics". *Open Education,* Vol. 2, (2005) pp. 43–51.
[R12.35] Kolin K. K. "The Evolution of Informatics," *Information Technologies,* Vol. 1, 2005, pp. 2–16.
[R12.36] Kolin K. K., "The Formation of Informatics as Basic Science and Complex Scientific Problems," In K.Kolin (Ed.), *Systems and Means of Informatics. Special Issue. Scientific and Methodological Problems of Informatics.* Moscow: IPI RAS, 2006, pp. 7–57.
[R12.37] Kolin K. K., Fundamental Studies in Informatics: A General Analysis, Trends and Prospects. *Scientific and Technical Information,* Vol. 1, (7), 2007, pp. 5–11.
[R12.38] Kolin K. K. "Structure of Reality and the Phenomenon of Information," *Open Education,* Vol. 5, 2008, pp. 56–61.
[R12.39] Losee, R. M., "A Discipline Independent Definition of Information," *Journal of the American Society for Information Science* Vol. 48(3), 1997, pp. 254–269.
[R12.40] Lozinskii, E. "Information and evidence in logic systems," *Journal of Experimental and Theoretical Artificial Intelligence*, Vol. 6, 1994, pp. 163–193.
[R12.41] Machlup, F., Mansfield, U. (eds.), *The Study of Information: Interdisciplinary Messages*, New York, Wiley, 1983.
[R12.42] MacKay, D. M., *Information, Mechanism and Meaning*, Cambridge, MIT Press, 1969.
[R12.43] Marr, D., Vision: *A Computational Investigation into the Human Representation and Processing of Visual Information.* San Francisco, W.H. Freeman, 1982.
[R12.44] Mingers, J., "The Nature of Information and Its Relationship to Meaning," In: Winder, R. L., et al. (eds.) *Philosophical Aspects of Information Systems*, London, Taylor and Francis, 1997, pp. 73–84.
[R12.45] Nauta, D., *The Meaning of Information*, The Hague, Mouton, 1972.
[R12.46] Newell, A. "The Knowledge Level". *Artificial Intelligence* Vol. 18, 1982, pp. 87–127.
[R12.47] Newell, A., Simon, H. A., "Computer Science as Empirical Inquiry: Symbols and Search," *Communications of the ACM*, Vol. 19, 1976, pp. 113–126.
[R12.48] Pierce, J. R., *An Introduction to Information Theory: Symbols, Signals and Noise*, New York, Dover Publications, 1980.
[R12.49] Poli, R., "The Basic Problem of the Theory of Levels of Reality" *Axiomathes*, Vol. 12, 2001, pp. 261–283.
[R12.50] Sayre, K. M.: *Cybernetics and the Philosophy of Mind.* London, Routledge and Kegan Paul, (1976).
[R12.51] Simon, H. A., *The Sciences of the Artificial*, Cambridge, MIT Press, 1996.
[R12.52] Smokler, H., "Informational Content: A Problem of Definition," *The Journal of Philosophy,* Vol. 63(8), 1966, pp. 201–211.
[R12.53] Ursul A. D., *The Nature of the Information. Philosophical Essay.* Moscow: Politizdat, (1968).
[R12.54] Ursul A. D., *Information. Methodological Aspects.* Moscow: Nauka, 1971.
[R12.55] Ursul A. D., *Reflection and Information.* Moscow: Nauka (1973).
[R12.56] Ursul A. D. *The Problem of Information in Modern Science: Philosophical Essays.* Moscow: Nauka (1975).
[R12.57] URSUL, A. D. "The Problem of the Objectivity of Information" in Iibor Kubátř, and Jiří Zeman (eds.), *Entropy and Information in Science and Philosophy,* New York, Elsevier, 1975.

[R12.58] Weaver, W., "The Mathematics of Communication," *Scientific American*, Vol. 181(1), 1949, pp. 11–15.
[R12.59] Winder, R. L., Probert, S. K., Beeson, I. A.: *Philosophical Aspects of Information Systems*, London, Taylor & Francis 1997.

R13. Planning, Prescriptive Science and Information in Cost-Benefit Analysis Analytics

[R13.1] Alexander Ernest R., *Approaches to Planning*, Philadelphia, Pa. Gordon and Breach, 1992.
[R13.2] Bailey, J., *Social Theory for Planning*, London, Routledge and Kegan Paul, 1975.
[R13.3] Burchell R. W. and G. Sternlieb (eds.), *Planning Theory in the 1980's: A Search for Future Directions*, New Brunswick, N.J., Rutgers University Center for Urban and Policy Research, 1978.
[R13.4] Camhis, Marios, *Planning Theory and Philosophy*, London, Tavistock Publication, 1979.
[R13.5] Chadwick, G., *A Systems View of Planning*, Oxford, Pergamon, 1971.
[R13.6] Cooke, P., *Theories of Planning and Special Development*, London, Hutchinson, 1983.
[R13.7] Dompere, Kofi K., and Taresa Lawrence, "Planning," in Syed B Hussain, *Encyclopedia of Capitalism*, Vol. II, New York, Facts On File, Inc., 2004, pp. 649–653.
[R13.8] Dompere, Kofi K., *Social Goal-Objective Formation, Democracy and National Interest: A Theory of Political Economy under Fuzzy Rationality*, (Studies in Systems, Decision and Control, Vol. 4), New York, Springer, 2014.
[R13.9] Dompere, Kofi K., *Fuzziness, Democracy Control and Collective Decision-Choice System: A Theory on Political Economy of Rent-Seeking and Profit-Harvesting*, (Studies In Systems, Decision and Control, Vol. 5), New York, Springer, 2014.
[R13.10] Dompere, Kofi K., *The Theory of Aggregate Investment in Closed Economic Systems*, Westport, CT, Greenwood Press, 1999.
[R13.11] Dompere, Kofi K., *The Theory of Aggregate Investment and Output Dynamics in Open Economic Systems*, Westport, CT, Greenwood Press, 1999.
[R13.12] Faludi, A., *Planning Theory*, Oxford, Pergamon, 1973.
[R13.13] Faludi, A. (ed.), *A Reader in Planning Theory*, Oxford, Pergamon, 1973.
[R13.14] Harwood, E. C. (ed.), Reconstruction of Economics, American Institute For Economic Research, Great Barrington, Mass, 1955., Also in John Dewey and Arthur Bently, 'Knowing and the known', Boston, Beacon Press, 1949, p. 269.
[R13.15] Kickert, W. J. M., *Organization of Decision-Making A Systems-Theoretic Approach*, New York, North-Holland, 1980.
[R13.16] Knight, Frank H. *Risk, Uncertainty and Profit*, Chicago, University of Chicago Press, 1971.
[R13.17] Knight, Frank H. *On History and Method of Economics*, Chicago, University of Chicago Press, 1971.

R14. Possible-Actual Worlds and Information Analytics

[R14.1] Adams, Robert M., "Theories of Actuality,"*Noûs*, Vol. 8, 1974, pp. 211–231.
[R14.2] Allen, Sture (ed.) *Possible Worlds in Humanities, Arts and Sciences*, Proceedings of Nobel Symposium, Vol. 65, New York, Walter de Gruyter Pub., 1989.
[R14.3] Armstrong, D. M., *A Combinatorial Theory of Possibility*. Cambridge University Press, 1989.
[R14.4] Armstrong, D. M., *A World of States of Affairs*, Cambridge, Cambridge University Press, 1997.

[R14.5] Bell, J. S., "Six Possible Worlds of Quantum Mechanics" in Allen, Sture (Ed.) *Possible Worlds in Humanities, Arts and Sciences*, Proceedings of Nobel Symposium, Vol. 65, New York, Walter de Gruyter Pub., 1989, pp. 359–373.

[R14.6] Bigelow, John. "Possible Worlds Foundations for Probability", *Journal of Philosophical Logic,* 5 (1976), pp. 299–320.

[R14.7] Bradley, Reymond and Norman Swartz, *Possible World: An Introduction to Logic and its Philosophy,* Oxford, Bail Blackwell, 1997.

[R14.8] Castañeda, H.-N. "Thinking and the Structure of the World", *Philosophia*, 4 (1974), pp. 3–40.

[R14.9] Chihara, Charles S. *The Worlds of Possibility: Modal Realism and the Semantics of Modal Logic*, Clarendon, 1998.

[R14.10] Chisholm, Roderick. "Identity through Possible Worlds: Some Questions", *Noûs,* 1 1967, pp. 1–8; reprinted in Loux, *The Possible and the Actual.*

[R14.11] Divers, John, *Possible Worlds*, London: Routledge, 2002.

[R14.12] Forrest, Peter. "Occam's Razor and Possible Worlds", *Monist,* 65 (1982), pp. 456–464.

[R14.13] Forrest, Peter, and Armstrong, D. M. "An Argument Against David Lewis' Theory of Possible Worlds", *Australasian Journal of Philosophy,* 62 (1984), pp. 164–168.

[R14.14] Grim, Patrick, "There is No Set of All Truths", *Analysis*, Vol. 46, 1986, pp. 186–191.

[R14.15] Heller, Mark. "Five Layers of Interpretation for Possible Worlds", *Philosophical Studies*, 90 (1998), pp. 205–214.

[R14.16] Herrick, Paul, *The Many Worlds of Logic,* Oxford: Oxford University Press, 1999.

[R14.17] Krips, H. "Irreducible Probabilities and Indeterminism", *Journal of Philosophical Logic*, Vol. 18, 1989, pp. 155–172.

[R14.18] Kuhn, Thomas S., "Possible Worlds in History of Science" in Allen, Sture (ed.) *Possible Worlds in Humanities, Arts and Sciences,* Proceedings of Nobel Symposium, Vol. 65, New York, Walter de Gruyter Pub., 1989. pp. 9–41.

[R14.19] Kuratowski, K. and Mostowski, A. *Set Theory: With an Introduction to Descriptive Set Theory*, New York: North-Holland, 1976.

[R14.20] Lewis, David, *On the Plurality of Worlds*, Oxford, Basil Blackwell, 1986.

[R10.21] Loux, Michael J. (ed.) *The Possible and the Actual*: *Readings in the Metaphysics of Modality,* Ithaca & London: Cornell University Press, 1979.

[R14.22] Parsons, Terence, *Nonexistent Objects,* New Haven, Yale University Press, 1980.

[R14.23] Perry, John, "From Worlds to Situations", *Journal of Philosophical Logic*, Vol. 15, 1986, pp. 83–107.

[R14.24] Rescher, Nicholas and Brandom, Robert. *The Logic of Inconsistency: A Study in Non-Standard Possible-World Semantics And Ontology*, Rowman and Littlefield, 1979.

[R14.25] Skyrms, Brian. "Possible Worlds, Physics and Metaphysics", *Philosophical Studies,* Vol. 30, 1976, pp. 323–332.

[R14.26] Stalmaker, Robert C. "Possible World", *Noûs*, Vol. 10, 1976, pp. 65–75.

[R14.27] Quine, W. V. O. *Word and Object*, M.I.T. Press, 1960.

[R14.28] Quine, W. V. O. "Ontological Relativity", *Journal of Philosophy*, 65 (1968), pp. 185–212.

R15. Rationality, Information, Games, Conflicts and Exact Reasoning

[R15.1] Border, Kim, *Fixed Point Theorems with Applications to Economics and Game Theory,* Cambridge, Cambridge University Press, 1985.

[R15.2] Brandenburger, Adam, "Knowledge and Equilibrium Games," *Journal of Economic Perspectives*, Vol. 6, 1992, pp. 83–102.

[R15.3] Campbell, Richmond and Lanning Sowden, *Paradoxes of Rationality and Cooperation: Prisoner's Dilemma and Newcomb's Problem*, Vancouver, University of British Columbia Press, 1985.

[R15.4] Gates Scott and Brian Humes, *Games, Information, and Politics: Applying Game Theoretic Models to Political Science*, Ann Arbor, University of Michigan Press, 1996.

[R15.5] Gjesdal, Froystein, "Information and Incentives: The Agency Information Problem," *Review of Economic Studies*, Vol. 49, 1982, pp. 373–390.

[R15.6] Harsanyi, John, "Games with Incomplete Information Played by 'Bayesian' Players I: The Basic Model," *Management Science,* Vol. 14, 1967, pp. 159–182.

[R15.7] Harsanyi, John, "Games with Incomplete Information Played by 'Bayesian' Players II: Bayesian Equilibrium Points," *Management Science*, Vol. 14, 1968, pp. 320–334.

[R15.8] Harsanyi, John, "Games with Incomplete Information Played by 'Bayesian' Players III: The Basic Probability Distribution of the Game," *Management Science*, Vol. 14, 1968, pp. 486–502.

[R15.9] Harsanyi, John, *Rational Behavior and Bargaining Equilibrium in Games and Social Situations*, New York Cambridge University Press, 1977.

[R15.10] Krasovskii, N. N. and A. I. Subbotin, *Game-theoretical Control Problems*, New York, Springer, 1988.

[R15.11] Lagunov, V. N., *Introduction to Differential Games and Control Theory*, Berlin, Heldermann Verlag, 1985.

[R15.12] Maynard Smith, John, *Evolution and the Theory of Games,* Cambridge, Cambridge University Press, 1982.

[R15.13] Myerson, Roger, *Game Theory: Analysis of Conflict*, Cambridge, Mass. Harvard University Press, 1991.

[R15.14] Rapoport, Anatol and Albert Chammah, *Prisoner's Dilemma: A Study in Conflict and Cooperation*, Ann Arbor, University of Michigan Press, 1965.

[R15.15] Roth, Alvin E., "The Economist as Engineer: Game Theory, Experimentation, and Computation as Tools for Design Economics," *Econometrica,* Vol. 70, 2002, pp. 1341–1378.

[R15.16] Shubik, Martin, *Game Theory in the Social Sciences: Concepts and Solutions*, Cambridge, Mass., MIT Press, 1982.

R16. Social Sciences, Mathematics and the Problems of Exact and Inexact Information

[R16.1] Ackoff, R. L., *Scientific Methods: Optimizing Applied Research Decisions*, New York, Wiley, 1962.

[R16.2] Angyal, A. "The Structure of Wholes," *Philosophy of Sciences*, Vol. 6, #1, 1939, pp. 23–37.

[R16.3] Bahm, A. J., "Organicism: The Philosophy of Interdependence" *International Philosophical Quarterly*, Vol. VII #2, 1967.

[R16.4] Bealer, George, *Quality and Concept*, Oxford, Clarendon Press, 1982.

[R16.5] Black, Max, *Critical Thinking*, Englewood Cliffs, N.J., Prentice-Hall, 1952.

[R16.6] Brewer, Marilynn B., and Barry E Collins (eds.) *Scientific Inquiry and Social Sciences,* San Francisco, Ca, Jossey-Bass Pub., 1981.

[R16.7] Campbell, D. T., "On the Conflicts Between Biological and Social Evolution and Between Psychology and Moral Tradition", *American Psychologist*, Vol. 30, 1975, pp. 1103–1126.

[R16.8] Churchman, C. W. and P. Ratoosh (eds.) *Measurement: Definitions and Theories*, New York, Wiley, 1959.

[R16.9] Foley, Duncan, "Problems versus Conflicts Economic Theory and Ideology" American Economic Association Papers and Proceedings, Vol. 65, May 1975, pp. 231–237.

[R16.10] Garfinkel, Alan, *Forms of Explanation: Structures of Inquiry in Social Science,* New Haven, Conn., Yale University Press, 1981.
[R16.11] Georgescu-Roegen, Nicholas, *Analytical Economics*, Cambridge, Harvard University Press, 1967.
[R16.12] Gillespie, C., *The Edge of Objectivity*, Princeton, Princeton University Press, 1960.
[R16.13] Hayek, F. A., *The Counter-Revolution of Science*, New York, Free Press of Glencoe Inc, 1952.
[R16.14] Laudan, L., *Progress and Its Problems: Towards a Theory of Scientific Growth,* Berkeley, CA, University of California Press, 1961.
[R16.15] Marx, Karl, *The Poverty of Philosophy,* New York, International Pub., 1971.
[R16.16] Phillips, Denis C., *Holistic Thought in Social Sciences*, Stanford, CA, Stanford University Press, 1976.
[R16.17] Popper, K., *Objective Knowledge*, Oxford, Oxford University Press, 1972.
[R16.18] Rashevsky, N. "Organismic Sets: Outline of a General Theory of Biological and Social Organism," *General Systems*, Vol. XII, 1967, pp. 21–28.
[R16.19] Roberts, Blaine, and Bob Holdren, *Theory of Social Process*, Ames, Iowa University Press, 1972.
[R16.20] Rudner, Richard S., *Philosophy of Social Sciences*, Englewood Cliff, N.J., Prentice Hall, 1966.
[R16.21] Simon, H. A., "The Structure of Ill-Structured Problems," *Artificial Intelligence,* Vol. 4, 1973, pp. 181–201.
[R16.22] Toulmin, S., *Foresight and understanding: An Enquiry into the Aims of Science,* New York, Harper and Row, 1961.
[R16.23] Winch, Peter, *The Idea of a Social Science*, New York, Humanities Press, 1958.

R17. Transformations, Decisions, Polarity, Duality and Conflict

[R17.1] Anovsky, 0mely M. E., *Linin and Modern Natural Science, Moscow*, Progress Pub., 1978.
[R17.2] Arrow, Kenneth J., "Limited Knowledge and Economic Analysis", American Economic Review, Vol. 64, 1974, pp. 1–10.
[R17.3] Berkeley, George, *Treatise Concerning the Principles of Human Knowledge, Works*, Vol. I (edited by A. Fraser), Oxford, Oxford University Press, 1871–1814.
[R17.4a] Berkeley, George, "Material Things are Experiences of Men or God" in [R1.5], 1967, pp. 658–668.
[R17.4b] Boulding, Kenneth E., *Ecodynamics: A New Theory of Societal Evolution,* Beverly Hills, Sage Pub., 1978, p. 11.
[R17.5] Brody, Baruch A. (ed.), *Readings in the Philosophy of Science*, Englewood Cliffs, NJ., Prentice-Hall Inc., 1970.
[R17.6] Brouwer, L. E. J., "Consciousness, Philosophy, and Mathematics," in Benecerraf, P. and H. Putnam (eds.), *Philosophy of Mathematics: Selected Readings*, Cambridge, Cambridge University Press, 1983, pp. 90–96.
[R17.7] Brown, B. and J Woods (eds.), *Logical Consequence; Rival Approaches and New Studies in exact Philosophy: Logic, Mathematics and Science*, Vol. II Oxford, Hermes, 2000.
[R17.8] Cornforth, Maurice, *Dialectical Materialism and Science*, New York, International Pub., 1960.
[R17.9] Cornforth, Maurice, *Materialism and Dialectical Method*, New York, International Pub., 1953.
[R17.10] Cornforth, Maurice, *Science and Idealism: an Examination of "Pure Empiricism"*, New York International Pub., 1947.
[R17.11] Cornforth, Maurice, *The Open Philosophy and the Open Society: A Reply to Dr. Karl Popper's Refutations of Marxism,* New York, International Pub., 1968.
[R17.12] Cornforth, Maurice, *The Theory of Knowledge*, New York, International Pub., 1960.

[R17.13] Dompere, Kofi K., "On Epistemology and Decision-Choice Rationality" in R. Trapple (ed.), *Cybernetics and System Research*, New York, North Holland, 1982, pp. 219–228.
[R17.14] Dompere, Kofi K. and M. Ejaz, *Epistemics of Development Economics: Toward a Methodological Critique and Unity*, Westport, CT, Greenwood Press, 1995.
[R17.15] Dompere, Kofi K., *The Theory of Categorial Conversion: Rational Foundations of Nkrumaism in socio-natural Systemicity and Complexity,* London, Adonis-Abbey Pubs., 2016–2017.
[R17.16] Dompere, Kofi K., *The Theory of Philosophical Consciencism: Practice Foundations of Nkrumaism in Social Systemicity,* London, Adonis-Abbey Pubs., 2016–2017.
[R17.17] Dompere, Kofi K., A General Theory of Information: Definitional Foundations and Critique of the Tradition, Working Monographs on Mathematics, Philosophy, Economic and Decision Theories, Washington, D.C., Department of Economics Howard University, 2016.
[R17.18] Dompere, Kofi K., *The Theory of Info-dynamics: Epistemic and Analytical Foundations,* Working Monographs on Mathematics, Philosophy, Economic and Decision Theories, Washington, D.C., Department of Economics, Howard University, 2016.
[R17.19] Dompere, Kofi K., *Polyrhythmicity: Foundations of African Philosophy*, London, Adonis and Abbey Pub., 2006.
[R17.20] Engels, Frederick, *Dialectics of Nature*, New York, International Pub., 1971.
[R17.21] Engels, Frederick, *Origin of the Family, Private Property and State*, New York, International Pub., 1971.
[R17.22] Ewing, A. C., "A Reaffirmation of Dualism" in [R1.5], pp. 454–461.
[R17.23] Fedoseyer, P. N. et al., *Philosophy in USSR: Problems of Dialectical Materialism*, Moscow, Progress Pub., 1977.
[R17.24] Kedrov, B. M., "On the Dialectics of Scientific Discovery," *Soviet Studies in Philosophy*, Vol. 6 1967, pp. 16–27.
[R17.25] Lenin, V. I. *Materialism and Empirio-Criticism: Critical Comments on Reactionary Philosophy*, New York, International Pub., 1970.
[R17.26] Lenin, V. I. *Collected Works Vol. 38: Philosophical Notebooks*, New York, International Pub., 1978.
[R17.27] Lenin, V. I., *On the National Liberation Movement*, Peking, Foreign Language Press, 1960.
[R17.28] Hegel, George, *Collected Works*, Berlin, Duncher und Humblot, 1832–1845 [also *Science of Logic*, translated by W. H. Johnston and L. G. Struther, London, 1951].
[R17.29] Hempel, Carl G. and P. Oppenheim, "Studies in the Logic of Explanation," in [R15.5], pp. 8–27.
[R17.30] Ilyenkov, E. V., *Dialectical Logic: Essays on its History and Theory*, Moscow, Progress Pub., 1977.
[R17.31] Keirstead, B. S., "The Conditions of Survival," American Economic Review, Vol. 40, #2, pp. 435–445.
[R17.32] Kühne, Karl, *Economics and Marxism, Vol. I: The Renaissance of the Marxian System*, New York, St. Martin's Press, 1979.
[R17.33] Kühne, Karl, *Economics and Marxism, Vol. II: The Dynamics of the Marxian System*, New York, St. Martin's Press, 1979.
[R17.34] March, J. C., "Bounded Rationality, Ambiguity and Engineering of Choice," *The Bell Journal of Economics*, Vol. 9 (2), 1978.
[R17.35] Marx, Karl, *Contribution to the Critique of Political Economy*, Chicago, Charles H. Kerr and Co., 1904.
[R17.36] Marx, Karl, *Economic and Philosophic Manuscripts of 1884,* Moscow, Progress Pub., 1967.
[R17.37] Marx, Karl, *The Poverty of Philosophy*, New York, International Publishers, 1963.
[R17.38] Marx, Karl, Economic and Philosophic Manuscripts of 1844, Moscow, Progress Pub., 1967.

[R17.39] Niebyl, Karl, H., "Modern Mathematics and Some Problems of Quantity, Quality and Motion in Economic Analysis," *Philosophy of Science*, Vol 7, #1, January, 1940, pp. 103–120.
[R17.40] Price, H. H., *Thinking and Experience*, London, Hutchinson, 1953.
[R17.41] Putman, H., *Reason, Truth and History*, Cambridge, Cambridge University Press, 1981.
[R17.42] Putman. H., *Realism and Reason*, Cambridge, Cambridge University Press, 1983.
[R17.43] Robinson, Joan, *Economic Philosophy*. New York, Anchor Books, 1962.
[R17.44] Robinson, Joan, *Freedom and Necessity: An Introduction to the Study of Society,* New York, Vintage Books, 1971.
[R17.45] Robinson, Joan, *Economic Heresies: Some Old-Fashioned Questions in Economic Theory,* New York, Basic Books, 1973.
[R17.46] Schumpeter, Joseph A., *The Theory of Economic Development*, Cambridge, Mass. Harvard University Press, 1934.
[R17.47] Schumpeter, Joseph A., *Capitalism, Socialism and Democracy*, New York, Harper & Row, 1950.
[R17.48] Schumpeter, Joseph A., "March to Socialism," *American Economic Review*, Vol. 40 May 1950, pp. 446–456.
[R17.49] Schumpeter, Joseph A., "Theoretical Problems of Economic Growth" *Journal of Economic History*, Vol. 8, Supplement 1947, pp. 1–9.
[R17.50] Schumpeter, Joseph A., "The Analysis of Economic Change," *Review of Economic Statistics*, Vol. 17, 1935, pp. 2–10.

R18. Vagueness, Approximation and Reasoning in the Information-Knowledge Process

[R18.1] Adams, E. W., and H. F. Levine, "On the Uncertainties Transmitted from Premises to Conclusions in deductive Inferences," *Synthese*, Vol. 30, 1975, pp. 429–460.
[R18.2] Arbib, M. A., *The Metaphorical Brain*, New York, McGraw-Hill, 1971.
[R18.3] Bečvář, Jiři, "Notes on Vagueness and Mathematics," in Skala, Heinz J. et al., (eds.), *Aspects of Vagueness*, Dordrecht, D. Reidel Co., 1984, pp. 1–11.
[R18.4] Black, M, "Vagueness: An Exercise in Logical Analysis," *Philosophy of Science*, Vol. 17, 1970, pp. 141–164.
[R18.5] Black, M. "Reasoning with Loose Concepts," *Dialogue*, Vol. 2, 1973, pp. 1–12.
[R18.6] Black, Max, *Language and Philosophy*, Ithaca, N.Y.: Cornell University Press, 1949.
[R18.7] Black, Max, *The Analysis of Rules*, in Black, Max [R18.8] *Models and Metaphors: Studies in Language and Philosophy*, Ithaca, New York: Cornell University Press 1962 pp. 95-139.
[R18.8] Black, Max, *Models and Metaphors: Studies in Language and Philosophy*, Ithaca, New York: Cornell University Press, 1962.
[R18.9] Black, Max *Margins of Precision*, Ithaca: Cornell University Press, 1970.
[R18.10] Boolos, G. S. and R. C. Jeffrey, *Computability and Logic,* New York, Cambridge University Press, 1989.
[R18.11] Cohen, P. R., *Heuristic Reasoning about uncertainty: An Artificial Intelligent Approach*, Boston, Pitman, 1985.
[R18.12] Darmstadter, H., "Better Theories," *Philosophy of Science*, Vol. 42, 1972, pp. 20–27.
[R18.13] Davis, M., *Computability and Unsolvability*, New York, McGraw-Hill, 1958.
[R18.14] Dummett, M., "Wang's Paradox," *Synthese,* Vol. 30, 1975, pp. 301–324.
[R18.15] Dummett, M., *Truth and Other Enigmas*, Cambridge, Mass. Harvard University Press, 1978.
[R18.16] Endicott, Timothy, *Vagueness in the Law*, Oxford, Oxford University Press, 2000.
[R18.17] Evans, Gareth, "Can there be Vague Objects?," *Analysis,* Vol. 38, 1978, p. 208.
[R18.18] Fine, Kit, "Vagueness, Truth and Logic," *Synthese,* Vol. 54, 1975, pp. 235–259.

[R18.19] Gale, S., "Inexactness, Fuzzy Sets and the Foundation of Behavioral Geography," *Geographical Analysis*, Vol. 4, #4, 1972, pp. 337–349.

[R18.20] Ginsberg, M. L. (ed.), *Readings in Non-monotonic Reason*, Los Altos, Ca., Morgan Kaufman, 1987.

[R18.21] Goguen, J. A., "The Logic of Inexact Concepts," *Synthese*, Vol. 19, 1968/69, pp. 325–373.

[R18.22] Grafe, W., "Differences in Individuation and Vagueness," in A. Hartkamper and H.-J. Schmidt, *Structure and Approximation in Physical Theories*, New York, Plenum Press, 1981. pp. 113–122.

[R18.23] Goguen, J. A, "The Logic of Inexact Concepts" *Synthese*, Vol. 19, 1968–1969.

[R18.24] Graff, Delia and Timothy (eds.), *Vagueness*, Aldershot, Ashgate Publishing, 2002.

[R18.25] A. Hartkämper and H. J. Schmidt (eds.), *Structure and Approximation in Physical Theories*, New York, Plenum Press, 1981.

[R18.26] Hersh, H. M. et al., "A Fuzzy Set Approach to Modifiers and Vagueness in Natural Language," *J. Experimental*, Vol. 105, 1976, pp. 254–276.

[R18.27] Hilpinen, R., "Approximate Truth and Truthlikeness," in M. Prelecki et al. (eds.) *Formal Methods in the Methodology of Empirical Sciences*, Wroclaw, Reidel, Dordrecht and Ossolineum, 1976 pp. 19–42.

[R18.28] Hockney D. et al. (eds.), *Contemporary Research in Philosophical Logic and Linguistic Semantics*, Dordrecht-Holland, Reidel Pub. Co., 1975.

[R18.29] Höhle Ulrich et al. (eds.), *Non-Classical Logics and their Applications to Fuzzy Subsets: A Handbook of the Mathematical Foundations of Fuzzy Set Theory*, Boston, Mass. Kluwer, 1995.

[R18.30] Katz, M., "Inexact Geometry," *Notre-Dame Journal of Formal Logic*, Vol. 21, 1980, pp. 521–535.

[R18.31] Katz, M., "Measures of Proximity and Dominance," *Proceedings of the Second World Conference on Mathematics at the Service of Man*, Universidad Politecnica de Las Palmas, 1982, pp. 370–377.

[R18.32] Katz, M., "The Logic of Approximation in Quantum Theory," *Journal of Philosophical Logic*, Vol. 11, 1982, pp. 215–228.

[R18.33] Keefe, Rosanna, *Theories of Vagueness*, Cambridge, Cambridge University Press, 2000.

[R18.34] Keefe, Rosanna and Peter Smith (eds.) *Vagueness: A Reader*, Cambridge, MIT Press, 1996.

[R18.35] Kling, R., "Fuzzy Planner: Reasoning with Inexact Concepts in a Procedural Problem-solving Language," *Jour. Cybernetics*, Vol. 3, 1973, pp. 1–16.

[R18.36] Kruse, R. E. et al., *Uncertainty and Vagueness in Knowledge Based Systems: Numerical Methods*, New York, Springer, 1991.

[R18.37] Ludwig, G., "Imprecision in Physics," in A. Hartkämper and H. J. Schmidt (eds.), *Structure and Approximation in Physical Theories*, New York, Plenum Press, 1981, pp. 7–19.

[R18.38] Kullback, S. and R. A. Leibler, "Information and Sufficiency," *Annals of Math. Statistics*, Vol. 22, 1951, pp. 79–86.

[R18.39] Lakoff, George, "Hedges: A Study in Meaning Criteria and Logic of Fuzzy Concepts," in, Hockney D. et al. (eds.), *Contemporary Research in Philosophical Logic and Linguistic Semantics*, Dordrecht-Holland, Reidel Pub. Co., 1975, pp. 221–271.

[R18.40] Lakoff, G., "Hedges: A Study in Meaning Criteria and the Logic of Fuzzy Concepts," *Jour. Philos. Logic,* Vol. 2, 1973, pp. 458–508.

[R18.41] Levi, I., *The Enterprise of Knowledge*, Cambridge, Mass. MIT Press, 1980.

[R18.42] Łucasiewicz, J., *Selected Works: Studies in the Logical Foundations of Mathematics,* Amsterdam, North-Holland, 1970.

[R18.43] Machina, K. F., "Truth, Belief and Vagueness," *Jour. Philos. Logic*, Vol. 5, 1976, pp. 47–77.

[R18.44] Menges, G., et al., "On the Problem of Vagueness in the Social Sciences," in Menges, G. (ed.), *Information, Inference and Decision*, D. Reidel Pub., Dordrecht, Holland, 1974, pp. 51–61.

[R18.45] Merricks, Trenton, "Varieties of Vagueness," *Philosophy and Phenomenological Research*, Vol. 53, 2001, pp. 145–157.

[R18.46] Mycielski, J., "On the Axiom of Determinateness," *Fund. Mathematics*, Vol. 53, 1964, pp. 205–224.

[R18.47] Mycielski, J., "On the Axiom of Determinateness II," *Fund. Mathematics*, Vol. 59, 1966, pp. 203–212.

[R18.48] Naess, A., "Towards a Theory of Interpretation and Preciseness," in L. Linsky (ed.) *Semantics and the Philosophy of Language*, Urbana, Ill. Univ. of Illinois Press, 1951.

[R18.49] Narens, Louis, "The Theory of Belief," *Journal of Mathematical Psychology*, Vol. 49, 2003, pp. 1–31.

[R18.50] Narens, Louis, "A Theory of Belief for Scientific Refutations," *Synthese*, Vol. 145, 2005, pp. 397–423.

[R18.51] Netto, A. B., "Fuzzy Classes," *Notices, Amar, Math. Society*, Vol. 68T-H28, 1968, p. 945.

[R18.52] Neurath, Otto et al. (eds.), *International Encyclopedia of Unified Science*, Vol. 1–10, Chicago, University of Chicago Press, 1955.

[R18.53] Niebyl, Karl, H., "Modern Mathematics and Some Problems of Quantity, Quality and Motion in Economic Analysis," *Science*, Vol 7, # 1, January, 1940, pp. 103–120.

[R18.54] Orlowska, E., "Representation of Vague Information," *Information Systems*, Vol. 13, #2, 1988, pp. 167–174.

[R18.55] Parrat, L. G., *Probability and Experimental Errors in Science*, New York, Wiley, 1961.

[R18.56] Raffman. D., "Vagueness and Context-sensitivity," *Philosophical Studies*, Vol. 81, 1996, pp. 175–192.

[R18.57] Reiss, S., *The Universe of Meaning*, New York, The Philosophical Library, 1953.

[R18.58] Russell, B., "Vagueness," *Australian Journal of Philosophy*, Vol. 1, 1923, pp. 84–92.

[R18.59] Russell, B., *An Inquiry into Meaning and Truth*, New York, Norton, 1940.

[R18.60] Shapiro, Stewart, *Vagueness in Context*, Oxford, Oxford University Press, 2006.

[R18.61] Skala, H. J. "Modelling Vagueness," in M. M. Gupta and E. Sanchez, *Fuzzy Information and Decision Processes*, Amsterdam North-Holland, 1982, pp. 101–109.

[R18.62] Skala, Heinz J. et al., (eds.), *Aspects of Vagueness*, Dordrecht, D. Reidel Co. 1984.

[R18.63] Sorensen, Roy, *Vagueness and Contradiction*, Oxford, Oxford University Press, 2001.

[R18.64] Tamburrini, G. and S. Termini, "Some Foundational Problems in Formalization of Vagueness," in M. M. Gupta et al (eds.), *Fuzzy Information and Decision Processes*, Amsterdam, North Holland, 1982, pp. 161–166.

[R18.65] Termini, S. "Aspects of Vagueness and Some Epistemological Problems Related to their Formalization," in Skala, Heinz J. et al., (eds.), *Aspects of Vagueness*, Dordrecht, D. Reidel Co. 1984, pp. 205–230.

[R18.66] Tikhonov, Andrey N. and Vasily Y. Arsenin., *Solutions of Ill-Posed Problems*, New York, Wiley, 1977.

[R18.67] Tversky, A. and D. Kahneman, "Judgments under Uncertainty: Heuristics and Biases," *Science*, Vil 185 September 1974, pp. 1124–1131.

[R18.68] Ursul, A. D., "The Problem of the Objectivity of Information," in Kubát, Libor and J. Zeman (eds.), *Entropy and Information*, Amsterdam, Elsevier, 1975, pp. 187–230.

[R18.69] Vardi, M. (ed.), *Proceedings of Second Conference on Theoretical Aspects of Reasoning about Knowledge*, Asiloman, Ca, Los Altos, Ca, Morgan Kaufman, 1988.

[R18.70] Verma, R. R., "Vagueness and the Principle of the Excluded Middle," *Mind*, Vol. 79, 1970, pp. 66–77.

[R18.71] Vetrov, A. A., "Mathematical Logic and Modern Formal Logic," *Soviet Studies in Philosophy*, Vol. 3, #1, 1964, pp. 24–33.

[R18.72] von Mises, Richard, *Probability, Statistics and Truth*, New York, Dover Pub., 1981.

[R18.73] Williamson, Timothy, *Vagueness,* London, Routledge, 1994.
[R18.74] Wiredu, J. E., "Truth as a Logical Constant with an Application to the Principle of the Excluded Middle," *Philos. Quart.*, Vol. 25, 1975, pp. 305–317.
[R18.75] Wright, C., "On Coherence of Vague Predicates," *Synthese*, Vol. 30, 1975. pp. 325–365.
[R18.76] Wright, Crispin, "The Epistemic Conception of Vagueness," *Southern Journal of Philosophy,* Vol. 33, Supplement, 1995, pp. 133–159.
[R18.77] Zadeh, L. A., A Theory of Commonsense Knowledge," in Skala, Heinz J. et al., (eds.), *Aspects of Vagueness*, Dordrecht, D. Reidel Co. 1984, pp. 257–295.
[R18.78] Zadeh, L. A., "The Concept of Linguistic Variable and its Application to Approximate reasoning," *Information Science*, Vol. 8, 1975, pp. 199–249 (Also in Vol. 9, pp. 40–80.

R19. Vagueness, Disinformation, Misinformation and Fuzzy Game Theory in Socio-Natural Transformations

[R19.1] Aubin, J. P. "Cooperative Fuzzy Games", Mathematics of Operations Research, Vol. 6, 1981, pp. 1–13.
[R19.2] Aubin, J. P. Mathematical Methods of Game and Economics Theory, New York, North Holland. 1979.
[R19.3] Butnaria, D., "Fuzzy Games: A description pf the concepts," Fuzzy sets and systems, Vol. 1, 1978, pp. 181–192.
[R19.4] Butnaria, D., " Stability and shapely value for a n – persons Fuzzy Games," Fuzzy sets and systems, Vol. 4, #1, 1980, pp. 63–72.
[R19.5] Nurmi, H., " A Fuzzy Solution to a Majority Voting Game, " Fuzzy sets and systems, Vol. 5, 1981, pp. 187–198.
[R19.6] Regade, R. K., "Fuzzy Games in the Analysis of Options," Jour. of Cybernetics, Vol. 6, 1976, pp. 213–221.
[R19.7] Spillman, B. et al., "Coalition Analysis with Fuzzy Sets," Kybernetes, Vol. 8, 1979, pp. 203–211.
[R19.8] Wernerfelt, B., "Semifuzzy Games" Fuzzy sets and systems, Vol. 19, 1986, pp. 21–28.

R20. Weapon Foundations for Information System

[R20.1] Forte, B., "On a System of Functional Equation in Information Theory," *Aequationes Math.* Vol. 5, 1970, pp. 202–211.
[R20.2] Gallick, James, *The Information: A History, a Theory, a Flood.* Pantheon, New York, NY, 2011.
[R20.3] Hopcroft, John, E. Rajeev Motwani and Ljeffrey D. Ullman, *Introduction to automata Theory, Languages, and Computation*, Pearson Education, 2000.
[R20.4] Howard, N., *Paradoxes of Rationality*, Cambridge, Mass., MIT Press, 1972.
[R20.5] Ingarden, R. S, "A Simplified Axiomatic Definition of Information," *Bull. Acad. Polo Sci. Ser, Sci. Math Astronomy Phys.*, Vol. 11, 1963, pp. 209–212.
[R20.6] Lee, P. M., "On the Axioms of Information Theory," *Annals of Math. Statistics*, Vol. 35, 1964, pp. 415–418.
[R20.7] Luce, R. D. (ed.), *Development in Mathematical Psychology*, Westport, Greenwood Press, 1960.
[R20.8] Luciano Floridi, "Is Information Meaningful Data?," *Philosophy and Phenomenological Research*, 70 (2), 2005 pp. 351–370.
[R20.9] Meyer, L, "Meaning in Music and Information Theory," *Journal of Aesthetics and Art Criticism,* Vol. 15, 1957, pp. 412–424.

[R20.10] Rich, Elaine, *Automata, Computability, and Complexity: Theory and Applications*, Pearson, 2008.

[R20.11] Shannon, Claude E., "The Mathematical Theory of Communication," *Bell System Technical Journal,* Vol. 27,#3, 1945, pp. 379–423 and Vol. 27, #4, 1948, pp. 623–666.

[R20.12] *Shannon, Claude E.* and Warren Weaver, *The Mathematical Theory of Communication.* University of Illinois Press, 1949.

[R20.13] Theil, Henri, *Statistical Decomposition Analysis*, Amsterdam, North-Holland, 1974.

[R20.14] Vigo, R. "Representational information: a new general notion and measure of Information". *Information Sciences.* Vol. 181 (2011), pp. 4847–4859.

[R20.15] Vigo, R "Complexity over Uncertainty in Generalized Representational Information Theory (GRIT): A Structure-Sensitive General Theory of Information". *Information.* **4** (1), 2013, pp. 1–30.

[R20.16] Vigo, R. *Mathematical Principles of Human Conceptual Behavior: The Structural Nature of Conceptual Representation and Processing*, Routledge, New York and London, 2014.

[R20.17] Wicker Stephen B. and Saejoon Kim, *Fundamentals of Codes, Graphs, and Iterative Decoding*. Springer, New York, 2003.

[R20.18] Wiener, N., Cybernetics, New York, Wiley, 1948.

[R20.19] Wiener, N., The Human use of Human Beings, Boston, Mass, Houghton, 1950.

[R20.20] Young, Paul. *The Nature of Information,* Greenwood Publishing Group, Westport, Ct, 1987.

R21. Weapon Foundations and Fuzzy Entropy

[R21.1] M. Belis, S. Guiasu, "A quantitative–qualitative measure of information in cybernetic systems," *IEEE Trans. Inform. Theory*, Vol. 14, 1968, pp. 593–594.

[R21.2] P. Burillo and H. Bustince, "Entropy on intuitionistic fuzzy sets and on interval-valued fuzzy sets," *Fuzzy Sets and Systems*, vol. 78, no. 3, pp. 305–316, 1996.

[R21.3] Ceng, H. D., Chen, Y. H. & Sun, Y. "A novel fuzzy entropy approach to image enhancement and thresholding," *Signal Processing, 75,* 1999, pp. 277–301.

[R21.4] De Luca, A. S. Termini, "A definition of non-probabilistic entropy in setting of fuzzy set theory", *Inform. Control.*, Vol. 20, 1972, pp. 301–312.

[R21.5] Dumitrescu, D., "Fuzzy measures and the entropy of fuzzy partitions," J. Math. Anal. Appl., Vol. 176, 1993, pp. 359–373.

[R21.6] Dumitrescu, D., "Entropy of fuzzy process," *Fuzzy Sets Syst.*, Vol. 55, 1993, pp. 169–177.

[R21.7] Dumitrescu, D., "Entropy of a fuzzy dynamical system," *Fuzzy Sets Syst.*, Vol. 70, 1995, pp. 45–57.

[R21.8] Garbaczewski, P., "Differential entropy and dynamics of uncertainty," *J. Stat. Phys.* Vol. 123, 2006, pp. 315–355.

[R21.9] Hu, Q. and Yu, D., "Entropies of fuzzy indiscernibility relation and its operations," *International Journal of Uncertainty, Fuzziness and Knowledge-Based Systems* Vol. 12, 2004, pp. 575–589.

[R21.10] Hu, Q., Yu, D., Xie, Z. and Liu, J. "Fuzzy probabilistic approximation spaces and their information measures," IEEE Trans. Fuzzy Syst., Vol. 14, 2006, pp. 191–201.

[R21.11] Hudetz, T. "Space-time dynamical entropy for quantum systems," Lett. Math. Phys., Vol. 16, 1988, pp. 151–161.

[R21.12] Hung, W. L. and M. S. Yang, "Fuzzy entropy on intuitionistic fuzzy sets," *International Journal of Intelligent Systems,* vol. 21, no. 4, 2006, pp. 443–451.

[R21.13] Kapur, J. N., "Four Families of Measures of Entropy," *Indian J. Pure Appl. Math.* Vol. 17, 1986, pp. 429–449.

[R21.14] Kapur, J. N. *Measures of Fuzzy Information*. Mathematical Sciences Trust Society, New Delhi, 1997.

[R21.15] Kasko, B. "Fuzzy Entropy and Conditioning," *Information Sciences,* Vol. 40, 1986, pp. 165–174.
[R21.16] G. J. Klir, "Generalized information theory: aims, results and open problems," *Reliab. Eng. Syst. Safety,* Vol. 85 (1-3), 2004, pp. 21–38.
[R21.17] Kolmogorov, A. N. *Foundations of the Theory of Probability*; Chelsea Publishing Company: New York, NY, 1950.
[R21.18] Kosko, B., "Fuzzy entropy and conditioning," *Inform. Sci.*, vol. 40, pp. 165–174, Dec. 1986.
[R21.19] B. Liu, Y. K. Liu, "Expected value of fuzzy variable and fuzzy expected value models," *IEEE Transactions on Fuzzy Systems,* Vol. 10, #4, 2002, pp. 445–450.
[R21.20] S. G. Loo, "Measures of fuzziness," *Cybernetica,* Vol. 20, 1977, pp. 201–210.
[R21.21] Markechová, D. "The entropy of fuzzy dynamical systems and generators," *Fuzzy Sets Syst.* 1992, Vol. 48, pp. 351–363.
[R21.22] Mesiar, R. and Rybárik, J. "Entropy of Fuzzy Partitions: A General Model," *Fuzzy Sets Syst.*, 99, 1998, pp. 73–79.
[R21.23] Mesiar, R. "The Bayes principle and the entropy on fuzzy probability spaces," Int. J. Gen. Syst., Vol. 20, 1991, pp. 67–72.
[R21.24] Parkash, O. "A new parametric measure of fuzzy entropy," *Inform. Process. Manage. Uncertain.* Vol. 2, 1998, pp. 1732–1737.
[R21.25] Parkash, O. and Sharma, P. K. "Measures of fuzzy entropy and their relations," *International Journal of Management & Systems,* Vol. 20, 2004, pp. 65–72.
[R21.26] Rahimi, M. and Riazi, A., "On local entropy of fuzzy partitions," *Fuzzy Sets Syst.*, Vol. 234, 2014, pp. 97–108.
[R21.27] Riěcan, B. "An entropy construction inspired by fuzzy sets," Soft Comput., Vol. 7, 2003, pp. 486–488.
[R21.28] E. Szmidt and J. Kacprzyk, "Entropy for intuitionistic fuzzy sets," *Fuzzy Sets and Systems,* vol. 118, no. 3, 2001, pp. 467–477.
[R21.29] Verma R. and B. D. Sharma, "On generalized exponential fuzzy entropy," *Engineering and Technology,* vol. 5, 2011, pp. 956–959.
[R21.30] Zeng W. and H. Li, "Relationship between similarity measure and entropy of interval valued fuzzy sets," *Fuzzy Sets and Systems,* vol. 157, #11, 2006 pp. 1477–1484.

R22. Written and Audio Languages and Information

[R22.1] Agha, Agha *Language and Social Relations,* Cambridge, Cambridge University Press, 2006.
[R22.2] Aitchison, Jean (ed.), *Language Change: Progress or Decay?* Cambridge, New York, Melbourne: Cambridge University Press, 2001.
[R22.3] Allerton, D. J. "Language as Form and Pattern: Grammar and its Categories". In Collinge, N.E.(ed.) *An Encyclopedia of Language.* London and New York: Routledge, 1989.
[R22.4] Anderson, Stephen, *Languages: A Very Short Introduction.* Oxford: Oxford University Press, 2012.
[R22.5] Aronoff, Mark; Fudeman, Kirsten, *What is Morphology.* New York, Wiley, 2011.
[R22.6] Barber Alex & Robert J Stainton (eds.), *Concise Encyclopedia of Philosophy of Language and Linguistics.* New York, Elsevier, 2010.
[R22.7] Bauer, Laurie (ed.), *Introducing linguistic morphology*, Washington, D.C., Georgetown University Press, 2003.
[R22.8] Brown, Keith; Ogilvie, Sarah, (eds.), *Concise Encyclopedia of Languages of the World,* New York, Elsevier Science, 2008.
[R22.9a] Campbell, Lyle (ed.), *Historical Linguistics: an Introduction,* Cambridge, MA: MIT Press, 2004.

[R22.9b] Chao, Yuen Ren, *Language and Symbolic Systems,* Cambridge, Cambridge University Press. 1968.

[R22.10] Chomsky, Noam, *Syntactic Structures.* The Hague: Mouton, 1957.

[R22.11] Chomsky, Noam, *The Architecture of Language.* Oxford: Oxford University Press, 2000.

[R22.12] Clarke, David S., *Sources of semiotic: readings with commentary from antiquity to the present.* Carbondale: Southern Illinois University Press, 1990.

[R22.13] Comrie, Bernard (ed.), *Language universals and linguistic typology: Syntax and morphology.* Oxford: Blackwell, 1989.

[R22.14] Comrie, Bernard, (ed.), *The World's Major Languages.* New York: Routledge, 2009.

[R22.15] Coulmas, Florian, *Writing Systems: An Introduction to Their Linguistic Analysis.* Cambridge University Press, 2002.

[R22.16] Croft, William, Cruse, D. Alan, *Cognitive Linguistics.* Cambridge, Cambridge University Press, 2004.

[R22.17] Crystal, David, *The Cambridge Encyclopedia of Language.* Cambridge: Cambridge University Press, 1997.

[R22.18] Deacon, Terrence, *The Symbolic Species: The Co-evolution of Language and the Brain,* New York: W.W. Norton & Company, 1997.

[R22.19] Devitt, Michael and Sterelny, Kim *Language and Reality: An Introduction to the Philosophy of Language.* Boston: MIT Press, 1999.

[R22.20] Evans, Nicholas; Levinson, Stephen C. "The myth of language universals: Language diversity and its importance for cognitive science". *Behavioral and Brain Sciences* **32** (5), 2009, pp. 429–492.

[R22.21] Fitch, W. Tecumseh, *The Evolution of Language.* Cambridge: Cambridge University Press, 2010.

[R22.22] Foley, William A., *Anthropological Linguistics: An Introduction,* Oxford, Blackwell, 1997.

[R22.23] Greenberg, Joseph, *Language Universals: With Special Reference to Feature Hierarchies.* The Hague: Mouton & Co., 1966.

[R22.24] Hauser, Marc D.; Chomsky, Noam; Fitch, W. Tecumseh, "The Faculty of Language: What Is It, Who Has It, and How Did It Evolve?," *Science 22*, **298,** (5598), 2002, pp. 1569–1579.

[R22.25] International Phonetic Association, *Handbook of the International Phonetic Association: A guide to the use of the International Phonetic Alphabet.* Cambridge: Cambridge University Press, 1999.

[R22.26] Katzner, Kenneth, *The Languages of the World,* New York: Routledge, 1999.

[R22.27] Labov, William, *Principles of Linguistic Change vol. I: Internal Factors,* Oxford, Blackwell, 1994.

[R22.28] Labov, William, *Principles of Linguistic Change vol. II: Social Factors,* Oxford, Blackwell, 2001.

[R22.29] Ladefoged, Peter, Maddieson, Ian, *The sounds of the world's languages*, Oxford: Blackwell, pp. 329–330, 1996.

[R22.30] Levinson, Stephen C. *Pragmatics.* Cambridge: Cambridge University Press, 1983.

[R22.31] Lewis, M. Paul (ed.), *Ethnologue: Languages of the World,* Dallas, Tex.: SIL International, 2009.

[R22.32] Lyons, John, *Language and Linguistics,* Cambridge, Cambridge University Press, 1981.

[R22.33] MacMahon, April M. S., *Understanding Language Change,* Cambridge, Cambridge University Press, 1994.

[R22.34] Matras, Yaron; Bakker, Peter, (eds). *The Mixed Language Debate: Theoretical and Empirical Advances.* Berlin: Walter de Gruyter 2003.

[R22.35] Moseley, Christopher (ed.), *Atlas of the World's Languages in Danger,* Paris: UNESCO Publishing, 2010.

[R22.36] Nerlich, B. "History of pragmatics". In L. Cummings (ed.), *The Pragmatics Encyclopedia.* New York: Routledge, 2010, pp. 192–193.

[R22.37] Newmeyer, Frederick J., *The History of Linguistics*, Linguistic Society of America, 2005.
[R22.38] Newmeyer, Frederick J., *Language Form and Language Function* (PDF). Cambridge, MA: MIT Press, 1998.
[R22.39] Nichols, Johanna *Linguistic diversity in space and time.* Chicago: University of Chicago Press, 1992.
[R22.40] Nichols, Johanna "Functional Theories of Grammar". *Annual Review of Anthropology,* Vol. **13,** 1984, pp. 97–117.
[R22.38] Senft, Gunter, *Systems of Nominal Classification.* Cambridge University Press. (ed.), 2008.
[R22.39] Swadesh, Morris, "The phonemic principle", *Language,* Vol. **10** (2): (1934), 117–129.
[R22.40] Tomasello, Michael "The Cultural Roots of Language". In B. Velichkovsky and D. Rumbaugh (Eds.), *Communicating Meaning: The Evolution and Development of Language.* Psychology Press, 1996, pp. 275–308.
[R22.41] Tomasello, Michael, *Origin of Human Communication.* MIT Press, 2008.
[R22.42] Ulbaek, Ib, "The Origin of Language and Cognition", In J. R. Hurford & C. Knight (eds.). *Approaches to the evolution of language.* Cambridge University Press. 1998, pp. 30–43.

Index

K.K. Dompere, *The Theory of Info-Dynamics: Rational Foundations of Information-Knowledge Dynamics*, Studies in Systems, Decision and Control 114, https://doi.org/10.1007/978-3-319-63853-9

R

S

T